Growing Blueberries

A Guide For The Small Commercial Grower

Published by
Rush River Publications
W4098 200th Ave.
Maiden Rock, WI 54750

Library of Congress Catalog Card No.: 96-______

ISBN: 1-888863-00-5

Table of Contents

Forward

Hello, and welcome to Rush River Produce, one of the finest places in the midwest to pick your own harvest of the best blueberries you can find. We are a fun day in the country for city folks and a great place to get lots of blueberries for the serious berry picker. We have great scenic vistas and perennial flower gardens that are at their peak of display during our blueberry season. We have been growing blueberries for ten years and are still learning how to grow and market them.

This book is written from personal experience combined with a lot of book research and conversations with other berry growers. It is based on real mistakes and successes. Please remember that this is not a "cookbook" for starting and running a blueberry farm. What has worked for us may or may not work in your situation. Please read this book as our "story", then you must do the work to find your own. Good luck.

If you are interested in developing a solid annual income growing great blueberries for the fresh market, this book will help you. This is not fast or easy money. You will need to invest a minimum of $3,000 to $5,000 annually (for each 1/2 to 1 acre), with a lot of sweat equity, and *you* **will not see any *return at all for 3 or 4 years***. You may not even know for sure that it will work for 2 or 3 years. You may not make a profit on your overall investment for 10 years or longer.

Our berry farm is called Rush River Produce. We bought our farm and started growing berries in 1986. We opened and started selling berries five years later in 1991. In year 10 our annual return on our total cash investment (not including our labor) was around 15%. In the next 5 years the return on overall investment should increase at a rate of 2% to 4% per year as blueberry bushes mature, marketing improves, and we get better at reducing losses to birds and other pests. We expect eventually to get an annual rate of return in the area of 20% to 30% on our overall actual cash outlay.

Actual net cash returns on established, mature, well managed blueberries can run from a few thousand dollars to $10,000 or more **per acre** each year. Crops vary from year to year. Weather and marketing affect production and sales. You will get better at managing your blueberry patch every year. This rate of return can be maintained for decades as a well managed blueberry patch may be productive for 35 to 50 years. An established planting requires a minimum of maintenance, e.g. weed control, fertilizer, irrigation, pruning, and pest control. The major investments are made in the five to eight years spent establishing the patch, bringing it into production, and developing your market. If you are still interested in raising blueberries please continue.

An established blueberry planting will require some maintenance work on your part, but unless you get really large, in excess of ten acres, it is not a full time job throughout the year. From January until March or April you should be spending a day or two a week planning, ordering tools, chemicals, replacement plants, and going to meetings to market your produce and learn more about the production and marketing end of the business. You should also plan on a day or so every week, whenever the weather is decent, pruning your mature blueberry plants. From April to the end of June you will have a few days of field work each week. You should also dedicate a few days to setting up advertising, and other marketing activities. This part of the year will approach a full time job for one person, but split between two people it will be two half-time jobs. If you are doing a U-pick, your sales season will, and must, become a madhouse for the duration of the picking season. We work 16 hour days from the middle of July until the blueberries are gone. This is currently about 25 days. In another 2 or 3 years this will cover about 3 months, from mid July until mid September, as our recently planted, later bearing blueberry varieties mature. October and November should be spent resting and doing residual weed control as soon as the blueberry plants are dormant.

Pruning the mature blueberry plants should start a few weeks after the plants go dormant. Given two people working on the overall project there will be two half time jobs from April through June and two **double-time** jobs for the berry season (1 to 3 months depending on varieties planted). There will be two half time jobs after the season for two months, and December through March should be a day or two a week planning, learning, and talking, and a day or two a week pruning. Your blueberry patch will rarely be a 9 to 5 job!

We can't find the money for you, and we aren't inclined to do the work for you, but we can pass on what we have learned in the process of actually building what has become one of the largest blueberry patches in the upper midwest. We want to help you develop a realistic idea of the cash and sweat investment required. We also hope to give you a pragmatic view of the time frame you can expect between startup and cash flow. Some of this information will be the dry technical information that you absolutely must know to start and run a successful blueberry patch. Some of the information will be our personal experiences in developing our blueberry patch.

Regardless of the quality of this information, and we think it reliable, you will need to do your own research. Start with the resources and materials listed at the end of the book. (If you know of some good information sources not listed here, let us know. We'll add them to the next edition.) Some of our experience will not apply directly to your situation. You will need to develop skills for digging out your own information, modifying ideas and practices to fit your situation, and occasionally just plain sticking your neck out to try something new.

We will also mention a few of our big mistakes - we learned to call them "educational experiences." If this book helps you in any way avoid making just one of these mistakes, it will have paid for itself hundreds of times over. We will start with big mistake number one:

BIG MISTAKE # 1: Don't try to do too many different projects, or start too big. If you have commercial horticultural experience you probably have a realistic idea of what can be accomplished by one or two people who have full time jobs on the side. If you are just getting started in the small fruit business you should get into it slowly. Depending on your level of experience and the amount of capital you have to invest, start with 1/2 to 1 acre of blueberries the first year. This will allow you to learn the ropes. (You will probably make a few mistakes in your first few years of operation, you don't want those mistakes to be too expensive.) Then you can realistically assess what you want to accomplish in subsequent years. Our first year we planted about 2/3 acre of blueberries, a few hundred apple trees, and a few acres of nut trees. We did it all again the next year. None of the nut trees and only a few of the apple trees are still alive. Remarkably, the blueberries lived through the neglect and are producing berries now. However, berry production and cash flow were delayed by the neglect. **Concentrate your efforts on one major project**. Preferably one with a good rate of return.

On the other hand we do believe strongly in "experimental" projects. We are constantly trying out new, and often unusual, crops that may prove profitable at some point in the future. I try to keep this "research and development" to less than 20% of our total effort. Concentrate your initial efforts on getting the blueberries to maturity. Then you should have time, and resources, to indulge other interests.

Introduction

Blueberries are one of the most popular small fruits grown in North America. I have met a few people who don't like them but not enough to count on both hands. The market for fresh blueberries is well established and quite large - with commercial production spread from Ontario, Canada, to Florida, and from British Columbia to Texas. Chile is already shipping "out of season" blueberries to northern markets. Millions of pounds of this delicious fruit are harvested from wild stands of low bush blueberries in Maine and eastern Canada for the processing market. More are grown and harvested from commercial highbush plantations in Michigan and other areas of the Midwest, the Pacific Northwest, and the South, for the processing and fresh markets. Recent introductions of new hybrid varieties from University of Minnesota, University of Wisconsin, and University of Michigan, have greatly extended the range of commercial blueberry culture to the north and west of traditional commercial blueberry growing areas.

The real problem with current commercial production and marketing practices is that the "fresh" blueberries in the grocery store are usually overripe and bland. They have lost most of their distinctive undertones of tartness which gives the blueberry the fresh picked "wild" blueberry flavor that most consumers want. The reason is simple - the berries are mechanically harvested when perfectly ripe then sorted, stored, shipped, stored at a warehouse, thrown on a truck, delivered to the store, and then put out for the consumer - a process that takes at best a few days and often a week or longer. The result is a product most often past its prime, soft, and bland.

There is an opportunity here for the local grower or the entrepreneur who wants to put in a few acres or more of blueberries for a profitable U-Pick or for fresh local marketing at farmers markets and roadside stands. Blueberries have a reputation for being difficult to grow. However, if the right conditions are met, they can be grown successfully just about anywhere south of USDA agriculture zone 2 or in areas where the winter low temperature is warmer than -45F degrees. There are a few tricks to growing blueberries, many which you will learn from this book. But there is much more to learn and there are plenty of other good resources to learn from. We list the best resources we know of at the end of this book.

Blueberries are somewhat more expensive to establish than other berry and small fruit crops - but not excessively so. They will take longer to reach full production than strawberries or raspberries. They compare to starting an apple orchard in that a blueberry patch can take five to ten years to reach full production - so your original investment is going to be tied up for awhile. But an established blueberry planting, if well managed, can be productive for thirty years or longer (longer than a modern apple orchard). It has relatively low per acre annual input costs once established. And it has an excellent profit potential if you can develop your market successfully.

A word of warning: we are not "organic" growers. While we sympathize with the goal of reducing and/or eliminating the use of agricultural chemicals, both for environmental and expense reasons, we have not yet figured out how to accomplish this without slave labor. We use a minimal amount of chemical fertilizers, and use them in a very directed fashion. We also use several pre-emergent herbicides to control weeds in the berry patch. All our chemicals are applied prior to leaf out so there is an absolute minimum of chemical residue on the ripe berries eight to sixteen weeks later. We find our customers are often interested, and we are as concerned as they are about the quality of the fruit they pick.

Throughout this book you will find sections that detail our personal experiences in starting and operating a small blueberry U-pick operation. Since we will discuss both our successes and our mistakes, don't immediately assume that these sections contain good advice. Also, a lot of this information is very site specific so even the "good advice" may not directly apply to your situation.

Terry and I were working in California in the mid 1980's and decided that it was time to settle down, have some kids, and figure out what to do when we grew up. I wasn't too keen on a steady job and a mortgage and retirement, and Terry was insistent that she wanted to start her own business. We spent a lot of time discussing possibilities while driving around on tours of the wineries or on hot sunny afternoons while picking raspberries and peaches at the local California U-Pick farms. We knew we were headed back to the midwest to be closer to Terry's family and we would discuss our options while working in the garden. When we looked at what we were doing while we talked about the possibilities an answer presented itself - gardening or horticulture.

We had a friend from Taiwan who suggested growing ginseng which she would market for us. We spent a lot of time at Chinese herb shops in Chinatown, San Francisco, talking with exporters and sampling different teas and herbs. We also spent a lot of time working up cash flow statements and sample business plans for the ginseng enterprise. Another friend thought he could get us some financing. The ginseng idea never worked out, but the experience of doing the research on growing techniques, marketing, and the cash flow projections taught us very valuable lessons about how to look at a small agricultural business enterprise.

Then the California jobs ended and we didn't feel a strong urge to hang around for more than a few months. So we headed back to the Midwest with a **PLAN**. The **PLAN** consisted of buying a small farm with decent soil, 40 acres or more, with a house, not on a major highway but not too far from one, within commuting distance of a major job market, and then start a berry farm. Kind of vague - but functional.

We knew an area which was developing into a low key tourist area. We liked the lay of the land, we had friends there, and it was also midway, and about an hour's drive, from three major cities in the area. We spent two weeks doing nothing but looking at real estate once we got there. We looked at thirty five or forty pieces of land in two weeks while talking to over a dozen real estate agents. We actually found six or seven places that would have worked as a berry farm and that were in our price range. What we settled on (yes, we made an offer after two weeks in the market), was larger than we wanted, with an older house in decent shape. It was a little past our upper price range but otherwise it was better than perfect - **it felt right!** (OK - we got lucky!) We are located on a bluff overlooking a small river valley. A short walk will take you to a beautiful scenic view of Lake Pepin on the Mississippi River. The soil is as good as it gets in this area - which is pretty good. The County Extension agent said "you can grow anything there." We have a microclimate that often adds two or three weeks to our growing season. Our bluff top location sheds light frosts very well. It's not quite perfect, but it's a lot closer to perfect than we deserve.

We didn't know all this when we bought the place. We knew it was good land and we liked the feel of it. **We liked to be there**. The neighbors were nice and seemed supportive. It seemed like a nice place to raise kids and to have people come to visit. (Later we figured out that we were selling this feeling and the scenic view almost as much as we were selling berries.) Remember this: you need to like what you are doing and where you are, and communicate this to your customers in any business but **especially** if you are direct marketing your produce.

Finances and Getting Started

There will be an initial and sizable investment in land and equipment. Since we are living on the farm and it cost about what we would have to spend to rent a decent apartment in town, we haven't included the cost of buying the land in our overall investment cost. This is a "no no" if you are taking a strict accounting of project costs. But then most accountants, or other investors, wouldn't invest money in a project with no immediate payback. We look at this business as a long term investment in our future. We label our big mistakes as "educational expenses" and value our life style as much as the actual income.

Overall this farm operation is designed to provide an income stream, after ten to fifteen years, which will allow our family to live comfortably, experiment with new projects, learn lots of new stuff, travel during the off season, and make a contribution to the local community. We are not all the way there yet, but we are getting close. Actually we are hoping that sales of this book will get us "over the hump," supplementing our berry income for a few years until all the blueberries are mature.

If you are going to go out to purchase some land to start a blueberry patch there are a few things that you should take into consideration. You need a market, that is you need to be accessible to a large number of people when your blueberries are ripe. Don't forget about parking space if you are planning a U-pick. Farmstead layout is important. Also, you should try to find land with a naturally acidic soil, a good water supply for irrigation, and good air drainage. Stay away from alkaline soils, clay soils, and low spots that may be subject to frost. If you already own a place that does not meet all these requirements (and our place does not) you can still grow blueberries (as long as you can get adequate water to them) - it is just more work, more money, and a little riskier.

Purchasing equipment is not too difficult. A more or less complete set of start-up equipment should be available in most rural areas for $3,000 to $8,000. You do not want to buy anything new if you can help it and you won't need everything right at the start. Buy used, but decent, equipment and, if at all possible, buy with cash. You want a 25 to 50hp tractor (Ford 8N, Massey 35, International 360 or 460 industrial, etc.) with a loader (if you can afford it) and probably a wide front, a two bottom plow, a brush hog type mower, a field cultivator and/or disc, a cheap lawn spreader for granular fertilizer, a small tractor mount sprayer, and a rototiller. Check out auctions, talk to local farmers, and check the local paper. In the berry business it is very important to be able to cobble machinery together from parts and to make do with old equipment, or use a piece of equipment to accomplish something that the designers didn't have in mind. Buying new equipment on credit is an almost sure fire way of making life 10% to 15% more difficult.

A computer can be a useful tool for marketing your blueberries. We originally bought one to process mailing labels to send a reminder to past customers that berries are ripe and it is time to pick (our best and cheapest form of advertising). Our mailing list was up to 1,100 names and we were tired of hand writing so many labels. Later it proved very valuable in organizing and producing our press release program (see section on Press Coverage). We mailed out 900 press releases over a six month period and were written about in over 30 newspaper and magazine articles, 3 television news programs, and even a radio talk show. You won't need a high end computer for these tasks, our 486 is really bigger than would be required. Check around for a used machine.

You will need an additional $3,000 to $8,000 per year to prepare your site and start planting 1/2 to 1 acre of berries and maintaining what is already in the ground. We have averaged on the high end of that range each year. I would recommend that you try to get as much going as you can at the front end of the project. It will bring in cash flow that much sooner. But balance this against the fact that you will be doing a lot of learning in this stage of the process and it is cheaper to make small mistakes than big mistakes.

To plant an acre of blueberries will require 1,000 to 2,000 plants depending on variety and field layout. Plants will cost $2 to $5 each depending on variety, size and delivery costs. You need to figure in the cost of sulfur and fertilizer, weed control (chemical or manual), irrigation supplies, gas for the tractor, equipment maintenance and so on. It's not cheap!

Everyone's situation is different but my strong advice boils down to (1) don't invest more than you can afford to lose, and (2) don't borrow the money. Interest paid out is money you have to work for but don't get to keep. You will probably have difficulty finding a lender willing to lend money for an untried venture of this type anyway. We financed the development of Rush River Produce out of our income from outside jobs and recommend that you do the same. Terry and I both had decent jobs (we commuted 60 miles each way) and we don't have extravagant tastes, so we could pretty comfortably accomplish this. Of course this also meant that we were working two full time jobs. A "real" job for investment and living money and then coming home to work on the farm for six to 12 hours. Then the kids came along and that is another full time job. The message here is that this project will take extra effort and a lot of sweat. If you need some advice on getting your financial house in order check Appendix A at the back of this book for some suggestions.

The best way to finance this type of project is out of your income based cash flow. If you are carrying the typical American debt load I would suggest that you can finance this project to a certain extent by getting out of debt (with the possible exception of land payments) and using the interest payments you don't make anymore for investment in the blueberry growing project.

Spend extra time and/or money on doing your taxes. If at all possible get someone used to dealing with farm and small business taxes to help you. Since you will probably be working an outside job and developing your own business at the same time your actual tax liability will go down dramatically. If you haven't been in business before, this may surprise you. The IRS will be glad to give you a year end 20% to 30% discount on everything you buy for your business and other business related expenses.

Keep detailed and accurate account of business related expenses. Set up a spreadsheet, either on paper or on computer, to keep things straight.

Also, try to set up large annual or semi annual fixed payments such as insurance premiums, equipment payments, and taxes to come due at the end of your berry sales season. It is much easier to take care of these important responsibilities in the relatively cash rich period after your season is over. This applies to general operating expenses also. You can buy next years irrigation supplies, fertilizer, and things at this time also, if you have an appropriate storage space. You will often be short of cash as your season starts with money flowing out for advertising, signs, and supplies. Try not to get caught having to make a tough decision between irrigating your berries and advertising the upcoming season.

Profit and Loss

In the process of digging out information on buying and growing blueberries, you will find different estimates of the potential production of each particular variety or type of blueberry at maturity. Generally, current information indicates the half-high types will produce 3 to 10 lb. per plant and the highbush types will produce 10 to 20 lb. per plant at maturity (7 to 9 years old) under good management and normal field conditions. These production estimates are **the starting point** for getting a handle on how much money you can make growing blueberries. Remember that you will not be able to reach the high end of these production figures for at least six years after planting 2 year old blueberry plants. Also, remember that the high end of these production figures represent almost perfect weather, great growing conditions, perfect use of all berries when they are ripe, and no losses to birds, hungry pickers, or bad weather.

To get a realistic estimate of the total berry production from an acre of mature blueberries you need to know the plant population per acre and make an estimate of the productive potential of the particular variety of blueberry. For highbush blueberry plants the standard planting pattern is 5 to 6 feet apart in the rows, with 10 feet between rows (some growers plant 4 feet by 10 feet but it makes picking by hand more difficult). Each plant takes up about 50 square feet. Divide this into 43,560 square feet in an acre, and you get a plant population of around 870 plants per acre (43,560/50 = 871). Most highbush blueberry varieties will produce between 10 and 20 pounds per plant at maturity so each acre should produce 8,700 to 17,400 pounds of blueberries per acre. STOP GETTING SO EXCITED - THIS IS ONLY POTENTIAL PRODUCTION. Actually, if you start with decent soil, and do a reasonably good job of soil preparation, fertilization, and weed control, you should expect production, after 6 or 7 years, in the range of 10 to 12 pounds per plant, or around 8,000 to 10,000 pounds per acre for highbush blueberry varieties. As you get better at the production side you can expect your production to get higher.

The half-high blueberry varieties will produce less per plant, 3 to 10 pounds, but are planted 3.5 to 4 feet in the row, by 5 to 7 feet between rows allowing plant densities of 1,500 to 2,400 plants per acre. At 5 pounds blueberry production per plant an acre of half-high blueberries can produce 8,500 to 12,000 pounds of berries.

There is a big difference between what the plants will produce and what you will get paid for. The difference is called management and marketing losses. This paragraph will be a litany of losses, but don't get too depressed. First, your blueberries will be harvested before they are dead ripe and so you will not get the full weight of berries in the production estimates. Then the birds (see section on pests) will eat a significant percentage of your berries - 25% to 50%, pickers may eat 5% of your berries. If you look in the section on marketing you will see that you should be giving away about 5% to 10% of your berries, and 10% to 20% will fall on the ground or just not get picked. Some of your plants will die every year. If you actually sell 50% of the berries your plants produce you will be doing very well, and you will have lots of room for improvement.

If you finally manage to get 4,000 pounds of berries sold from each acre of mature blueberries you will be in pretty good shape. Depending on the local competition your U-pick selling price may range from $0.90 per pound (if your are near a commercial growing area and have a lot of competition), to $2.00 per pound (if you have no competition at all in a good market). Pre picked berries will go for $1.50 to $4.50 a pound depending on the market, but you will have increased labor cost for picking the berries. We sell nearly all our berries U-pick at $1.85 a pound. We have some competition in our market, but we are the largest blueberry grower in our area so people who want a lot of blueberries come here. Our prices are not the highest in the area but are toward the high end. You will need to spend some time looking at the local market, if there is one, and decide where you want to fit into it.

4,000 pounds times $0.90 is still $3,600 gross profit per acre. 4,000 pounds times $1.85 is $7,400 gross profit per acre. You will spend around $250 to $1,000 per acre on fertilizer, weed control chemicals, irrigation supplies, and equipment depreciation. You will also be spending $500 to $1,000 per acre on marketing and advertising. If you keep your operations and marketing expenses down to say $1,200 per acre, your net profit could run from $2,400 ($0.90 berries), to $6,200 ($1.85 berries) per acre. At these rates of return, 5 to 10 acres of blueberries will provide a basic living once all your plants are mature.

Remember this is a conservative estimate of what you can make when you start and operate a blueberry patch. If you pay attention to what you are doing, study production techniques, go to growers meetings, spend a little extra to improve you soils, you will realize more of the potential production of your blueberry plants. You should be able to improve your production by 2% to 3% each year. You should also be able to reduce your management losses by a similar amount each year by better marketing and new, improved, anti bird tactics.

If you are able, after several years of applied study and growing experience, to realize 80% of your blueberry plants **production potential** (say 14,000 pounds per acre), you will be doing very well. If, in addition to that you are able to keep your **management losses** down to 20% of production (say you actually sell 11,000 pounds of berries per acre), at $1.85 per pound that's $20,350 gross profit per acre. With expenses for operations and marketing of $4,000 per acre (It will cost more to produce and more to sell a larger amount of blueberries) you could hit a net profit of $16,350 per acre, when you get really good at growing and marketing. We are still learning and trying to get good at it. I estimate that we are actually selling about 60% of the berries we produce.

The Blueberry Plant

The blueberry is a member of the heath family, ***Ericaceae***, and is related to cranberries, lingonberries, rhododendrons, heaths, and huckleberries. This family of plants has several unique characteristics which distinguish it from much of the rest of the horticultural world.

The plants in the **Ericaceae** family appear to have evolved in areas where high annual rainfall and low soil calcium have resulted in soil pH (acidity) between 4.0 and 5.5. Often associated with swampy areas, blueberries do not actually grow well in areas with significant soil saturation. They are often found in highly organic soils on raised hummocks in or around swampy areas. They also occur in open or lightly wooded sandy soils where the water table is shallow. For optimum growth in a commercial setting you must learn to mimic the natural conditions where the blueberry developed and improve upon them wherever possible.

The blueberry has a relatively slow growing root system which lacks the fine root hairs typical of the roots of most plants. Roots tend to grow very near the soil surface with 80% of the roots in the top 12 inches of soil. To compensate for lack of surface area and absorptive capacity usually supplied by root hairs, the plants have developed an association, or symbiosis, with mycorrhizal fungi which functionally replace the root hairs. These fungi actually grow into the blueberry root structure and feed off carbohydrates stored in the blueberry root cells. The blueberry plants let the fungi get away with this because the fungi supply the plant with a greatly increased supply of water and mineral nutrients. In fact, the mycorrhizal fungi are so efficient at extracting minerals from soil that the blueberry plant is able to survive in very poor soils. If you attempt to raise blueberries on better soils you may never need any fertilizer beyond a yearly application of nitrogen.

Another atypical characteristic of blueberries is the structure of the flower. The flower is bell shaped, somewhat enclosed, with the open end away from the source of both nectar and pollen. The shape of the flowers make pollination by domestic honeybees difficult. The most efficient pollinators are bumblebees and other wild bee species. It is beneficial to have areas nearby the blueberry patch which are left wild as habitat for wild bee populations. Fallen logs, dead trees, grass land, stacks of hay, even old brush piles supply the necessary habitat for wild bees of different species. We also grow a few acres of fall bearing raspberries which are much frequented by wild bees through the fall and, I believe, supply a large late season food source which helps these critters over winter. The wild bees will be your star pollinators. However, just as insurance, you should plan on having 1 healthy hive of domestic honey bees for each acre of mature blueberries you want to pollinate. We generally rent 10 hives for our seven acres of blueberries from a local beekeeper at a nominal cost.

Our blueberry patch is located near old pasture and quite a large wooded area. There is a weedy wind break all along one edge of the patch. Every year or two I stack up 6 or 8 bales of hay inside the wind break in a few places. I have also pushed large branches and logs into the wind break. Sure, it looks messy, **but it is prime wild bee habitat**. If your patch is in a more developed area you will want to concentrate on developing wild bee habitat. Even if you live in a wilder, less developed area, if you plan on having a patch larger than an acre or two you will want to develop wild bee habitat as you develop your patch.

The blueberry, along with other members of the heath family, requires a constant liberal supply of water during the fruit production season. Unless your site is located in an area with a very shallow water table, within 18" to 24" of the surface, you will need to provide some form of irrigation on a regular basis.

Blueberry Varieties

There are basically four commercial types of blueberry plants available today. This is a functional description, not a botanical classification. Actually, many of these varieties are interrelated, and they show a wide set of differing characteristics. They have been extensively bred, selected, and hybridized over the past 100 years to develop the commercial blueberry varieties available today.

The Rabbiteye blueberry is a different species of plant than the commercial highbush varieties though there some hybrids. Rabbiteyes are varieties commonly grown in the Gulf Coast states. They have been selected and bred for areas with very mild winters. These varieties have a very low chilling requirement (approximately 150 to 500 hours below 45F) before they will break bud in the spring. They lack any significant cold hardiness. Some Rabbiteye varieties are more tolerant of less acidic soils than other types of blueberries.

Our experience is in growing highbush and half high blueberry varieties. Most of the information in this book is applicable to Rabbiteye blueberries but growing blueberries in the South is not our specialty. Prospective rabbiteye blueberry growers should check information in this book with someone having local experience. Also spend extra time reading the general blueberry texts in Appendix A, most of which contain extensive information on rabbiteye culture.

The commercial **Highbush blueberry** has been selected, bred, and hybridized for over 100 years and has an incredible range of adaptability. Recently varieties have been introduced which rival the Rabbiteye types in heat tolerance and adaptability to warm winter temperatures. On the other extreme is the variety Patriot which is reputed to be hardy and productive into USDA zone 3 and has withstood -35F at our place with little injury.

The **"Half high"** blueberry varieties, introduced mostly over the last ten to fifteen years, are generally the result of hybridization of commercial highbush varieties with the wild low bush blueberry. The result has been lower statured plants, from 18 inches to 4 feet tall, that are reliably cold hardy to -30F and colder, with some varieties as productive, on a per acre basis, as commercial highbush plants. Many varieties have been introduced in the past decade which extend reliable production of blueberries well into Zone 3 and possibly, in special situations, into colder areas. These varieties often depend on winter cover of either snow or floating row cover material for winter protection and maximum productivity.

The **wild low bush blueberries** are common in many areas of the country. They tend to have small berries, variable annual productivity, and, with the exception of commercial development of extensive naturally occurring stands in the North East U. S. and Eastern Canada, are not generally adapted to cultivation on a commercial scale.

Choosing varieties for growing in your area is not difficult. For U-Pick operations, or fresh market, you are looking for varieties having large berry size, ease of picking, high quality (good flavor, firmness, good color, etc.), and adaptability to your minimum winter temperatures. You should order catalogs from any and all suppliers your can locate (see Appendix B). The best sources of information about blueberry cultivars for your area will be a combination of information from suppliers' catalogues, your County Extension Agent and University Extension publications, other blueberry growers in your area (or other small fruit growers who may have tried blueberries) and other books and publications (see Appendix A).

If your live in an area where blueberries are not usually grown don't be surprised if your county agent gives you a blank look. He or she may not know much about blueberries but should be able to get you information from your State University system, or publications from other State Extension offices if you ask. Your Extension Agent can help you with some of the necessary research you should do but there is a lot of information out there. In the long run the responsibility is yours and the more information you have in hand when you start, the more likely you are to start off in the right direction.

Blueberry plants are currently propagated by two methods. **Tissue culture** plants are available from many suppliers and are generally sold as tissue culture plugs or smaller, 4" to 6" potted plants. The upside of tissue culture propagation is that there is little likelihood of introducing diseases to your planting by these plants, and tissue culture is a cheap way to propagate a large number of plants. The down side is that, occasionally, these plants are not well hardened off for field planting, and they are generally sold as younger, smaller plants. These smaller plants will often require intensive care in a nursery bed for at least a year prior to planting out in the field.

Field grown plants may be either tissue culture plants set out in a nursery row for a year or two, or plants propagated from cuttings in the nursery, and grown in the field for one to three years. While it is theoretically possible to introduce diseases to your planting from these sources, commercial nurseries are inspected regularly, and the actual likelihood of getting diseased plants is small. The field grown plants should be well hardened off, and two or three year old plants are easily large enough to field plant immediately. I have purchased two and three year old plants that actually produced ripe berries the year they were planted.

An additional benefit from purchasing field grown blueberry plants is the introduction of mycorrhiza, fungal organisms that are associated with the blueberry roots, which assist the plants in water and nutrient absorption. These fungal organisms may or may not be present in your soil, and may not be associated with plants obtained directly from tissue culture. Blueberry plants acquired from commercial blueberry nurseries are generally grown in soils which supply the correct fungal organisms.

BIG MISTAKE #2: Don't buy smaller, younger, or cheaper plants to save a few hundred dollars at the front end of the project. You will lose more plants and your planting will be later bearing. Buy the most robust 2 or 3 year old blueberry plants you can find.

One thing I would do differently, at the start of this project, would be to get a selection of plants from every supplier I could locate and compare them at the end of the first summer. Pick the two suppliers who send you the best plants and stick with them. You shouldn't work with just one supplier - they won't always have the varieties you want.

Terry and I got started in the direction of blueberry farming by a salesman from a local wholesale nursery. I was talking to him about apple trees and raspberries, and he asked, "How would you like to grow a small bush that produced a ten dollar bill every year?" He proceeded to explain about some recent releases of new blueberry varieties from the University of Minnesota: Northblue, Northsky, and Northcountry. We didn't know anything about blueberries except that they were uncommon in our area. We discovered that these varieties were new releases, and that a few growers were using them for small test plantings. We talked to people at the University of Minnesota, we talked to the people who propagated the plants, we went to the local fruit growers meetings, we talked to our County Agent, we talked to University of Wisconsin berry people (there's where we got the blank look I mentioned - but, to their credit, they said to let them know how it worked out - they were interested in our results).

What we found out from all this research was that these varieties were hardy enough for our area, but that our soil was **too good** for blueberries. We have a good silt loam soil that grows a lot of corn, clover, and alfalfa and has a nice, but only slightly acid, pH of 6.5 to 6.7. This is not normally considered to be a good blueberry soil and we certainly didn't live in an area where there was any significant blueberry production.

We also found out that it is possible to change soil pH and to maintain it at appropriate levels but that there would need to be continual soil tests and probably significant additional pH adjustments made in the following years to maintain the plants vigor and production. Now this obviously presents a series of problems and challenges, but it is also an extraordinary opportunity - if we could figure out how to do this correctly, or even just pretty well, we would have essentially no competition in the fresh blueberry market in our area.

We had considered growing strawberries, but every county in our area has three or four strawberry growers. The same holds true for raspberries. It's certainly possible to start up a business when there are other people in the same business in your area, but that does not make it easy. To be real frank - we knew that we didn't know as much about what we were going to be doing as the people already running more or less successful berry operations. So we figured the less competition we had the better off we were.

So we jumped in with both feet and planted nearly an acre of half high Northblue berry plants. The next year we planted another acre, and the next, and the next. After the fourth year of planting we were growing nearly 10% of all the Northblue plants that existed in the world. We didn't even know yet if they would ever fruit. When we planted our fourth acre of Northblue we were just starting to get a few handfuls of blueberries off of our first planting.

A note here about planting large numbers of plants when you don't know how to plant large numbers of plants. It's not really difficult -- what you really need is a lot of friends who have a sense of humor and are forgiving. We realized this early and planted our first acre of blueberries by **the party method**. We got everything ready as best as we knew how and invited our friends out from the city and put them to work for a few hours. Projects that you cannot do alone can be accomplished rapidly and much more efficiently when done by a large group of friends (who will also operate much more efficiently than hired temporary labor). A large group of people will always have suggestions for making the work more efficient and easier and you will learn a lot from them. **The Planting Party** is an institution now. Last year we planted 1,200 highbush blueberry plants by hand in about three hours and ten minutes. The plants went in faster than I could hook up the irrigation lines to water them.

When the planting is done we have a big picnic and everyone celebrates a big job well done with no real strain on anyone. The planting crew has gone from about twenty to around sixty through addition and multiplication (lots more kids now) in the past ten years. We have to turn down requests for invitations. The deal is that everyone who helps plant gets free berries for life and a sense of accomplishment that won't quit.

After a few years of planting mostly Northblue (we planted a few interspersed Northcountry for pollination) we realized that we were at a point (about 4 acres) where adding more of that variety would only cause us problems. Northblue ripens over a 21 to 26 day period in our area so when it is picking time life gets pretty chaotic around here. We realized that once these bushes were really mature, and producing five to ten pounds of berries each, that we could not get enough cars down our road in that time period to get all those berries picked. As much as we wanted to expand our blueberry plantings, unless we could find other varieties with a different ripening season, we were done planting blueberries. The University of Minnesota came out with St. Cloud, a newer half high variety but it was essentially the same season as Northblue, perhaps a few days earlier. The only other varieties that we knew of were the standard commercial varieties from Michigan. There were more of these varieties to choose from with a much wider range of ripening seasons and other characteristics. But we didn't think they would grow reliably this far north and west. We are on the Minnesota/Wisconsin border about an hour south and east of the Minneapolis/St. Paul area.

Then, one day right at the end of our second season of actually selling a few blueberries, two older guys showed up to pick blueberries. Their first reaction was that the berries were nice but the plants were pretty short. They asked why didn't we grow the regular big ones like the other fellow a hundred miles east of here. Who, What, Where, What's his phone number?

Five weeks later we drove up to an orchard about a hundred miles east of our place, similar soils, maybe a little colder, maybe not, and there are rows and rows of five to six foot tall commercial highbush blueberry bushes just loaded with bunches of blueberries - and this is September 1st, five weeks after **our** season is over. This guy and his family are growing regular commercial highbush blueberries, nine acres of them, because "nobody told us it wouldn't work." He was growing varieties that weren't even considered particularly hardy for commercial highbush varieties. We learned a lot that afternoon.

Terry and I went home and started working on our next blueberry order. We picked from among the hardier, large fruited commercial highbush varieties and tried to spread our season out as much as possible. Our choices were Bluecrop, the main commercial variety and Patriot, a very hardy and large fruited variety. Both start to ripen a little earlier than Northblue but ripen over six weeks instead of three. Nelson and Elliot are both late varieties that will start ripening as the Northblue finish. We planted three more acres of those four varieties over the next two years.

Winter of 1995-96 was a test year in our area. The coldest temperatures in 35 years took us down to -35F one night and close to that several other nights. Ironically those nights were spent at a fruit growers conference in St. Cloud, Minnesota. Patriot survived the cold snap with little detectable winter injury and produced enough berries during the '96 season to be worth picking on a regular basis. The Bluecrop, Elliot, and Nelson varieties all froze back to the snow line, about 18 inches. All these varieties showed vigorous growth during the next growing season, but produced few berries. This may actually benefit the plants in the long run. They were only in their second year in the ground and probably put on more vegetative growth this year than they might have if they had a significant berry crop. In any case these varieties seem to be recovering well and should produce a good crop next year if temperatures remain in a normal range this coming winter.

This coming season (1997) we will see the first production from the highbush plants - and next year the blueberry season should run from the beginning of July into September. Next year will see a significant increase in both volume and length of our season. In the years after that we should have blueberries from early to mid July well into September and the three acres of highbush Bluecrop, Patriot, Nelson and Elliot varieties should equal the production of the four acres of half high Northblue.

Propagation

Initially you should be more interested in getting your blueberry acreage planted than in propagation. Yes, it is possible, if you know what you are doing, to propagate your own plants once you have sufficient plant material to propagate from. I don't recommend doing this yourself at startup unless you are already in the plant propagation business. Professional blueberry nurseries can do it better and cheaper than you can.

You may want to consider propagating replacement plants once your blueberry planting is established and you are working at it full time. At that point propagating several hundred plants each year to replace dead or unhealthy blueberry plants will be a small diversification, not a major project.

Remember, many plant varieties, especially newer releases, are protected from unlimited propagation by patents and/or exclusive rights to propagation. You may, in many cases, request limited propagation rights from the patent holder. If you fail to get these rights, in writing, you may be subject to legal action by the owners of the plant's patents.

I will leave the details of the propagation process to the text books mentioned in Appendix A. The commercial nurseries use either tissue culture, or softwood cuttings for large scale propagation. On a smaller scale stooling is a satisfactory alternative, requiring less set up and labor.

Finding and Developing Your Market

You should research four questions before you start a blueberry patch. (1) Do people in your area buy fresh blueberries? (2) Do you have any competition? (3) Do you like working with people? (4) Are you willing to work very long hours during the berry season? If people in your area buy a lot of blueberries and you don't have any competition, you are in the best of all possible worlds. If they buy blueberries, and there is local competition your income will probably be lower because of competition and the extra expense of developing new customers. If people don't buy a lot of blueberries but there is no competition you have an excellent chance of doing well if you can learn how to market the berries.

Operating a successful berry patch is about one half horticultural skill and learning and one half marketing skill and learning. You should always plan on spending extra money on experimental marketing every year until you have enough customers and I don't know any berry growers who would admit to having enough customers. Experimental marketing is anything from advertising in a new outlet, or giving free coupons to local businesses to give to their customers, to having lots of extra Tiny Tim pumpkins to give away as treats to the kids. You should plan on spending as much as you take in from berry sales on advertising for the first several years you are open, probably more. You will spend several years starting your mailing list. The cheapest advertising you can do is letting your past customers know that berries are ripe. If they liked you and your berries they will return for the cost of a post card.

You must always track your marketing efforts by asking your customers where they heard about you. "How did you find out about us?" should be the second thing you say to them after welcoming them. This is the only way to find out what advertising is actually delivering customers to your door. If, as you talk with your customers, you find that some advertisements aren't bringing you results - drop the accounts. Don't waste time and money. On the other hand you should always have a part of your advertising budget earmarked for experimentation. Every year you should try out one or several new advertising outlets. Keep the best and drop the rest.

You will be spending most of your income, and then some, on advertising for the first few years you are in operation. We try to spend about $3,000 per year on advertising. New customers are expensive. We figure it costs us between $7 and $8 to get a new customer in the field to pick berries. That is not far below our average sale so we don't make any money on our first time customers. We have to bring them back, this season and next season, to make money on them.

I am a strong believer in giving away berries as a marketing tool. Every year we mail our advertising line cards, with a coupon for a pound of free berries, to every Bed and Breakfast within sixty miles. If a local service organization is holding a raffle we give them some coupons for 5 or 10 pounds of free berries to include in the raffle. We have offered 10 pounds of free berries to local groups for fundraising purposes, and we announce this through the local newspapers. We also donate some frozen berries to the local pancake breakfast for the ambulance fund. In one way or another we give away a lot of blueberries, perhaps 5% of our crop, for local goodwill. This generates lots of first time customers for us and it generates a lot of local goodwill. We have tried giving a "good pickers" discount - pick 10 pounds of berries and get the next pound free. So we actually gave away up to 10% of our berries for marketing purposes. We dropped that promotion after a one year trial. It was too expensive.

Getting "free ink" is another very important source of free advertising. Being featured in local magazines and newspapers is better than paid advertising because it gives you much better exposure and it makes you "realer" to most people. How do you get "free ink?" For the first round it should be enough to be a new business in your area doing something a little different. Local newspapers and shoppers are usually looking for local interest stories, but you may need to present yourself as a possibility. Don't expect them to come beating down your door on their own.

Our second year in business I called our county newspaper and asked if they would do a story about us as a new local business. The editor was happy to do it and drove out the next day. He did a real nice front page story, with a picture, all about the new blueberry farm in the area. I suspect that in the next week he got calls from every other berry grower in the county because the next issue had an article about all the berry growers in the county, including us. Then the newspaper in the town across the river must have thought that berries were big news because they did an article about all the berry growers in that county, and included us because we were the only blueberry grower in the area. After that the local weekly shopper decided that we were news. A few weeks later a regional agricultural magazine interviewed us and took some pictures. Another front page story!

Developing these opportunities in the local media is an art. We are trying to do this all ourselves so we try out lots of different things. Each year we send out press releases every month from April until the picking season is over in September. We have defined "local media" as any publication that might be remotely interested in us located within an 80 mile radius. We will let them know we are here in April, tell them about our lingonberry research project in May, tell them about apricots and black currants in June, and tell them about the blueberry season again the first of July, and so on. We have about 150 media outlets on the mailing list. If only 3% pick up on it every month it will mean 15 to 20 stories about Rush River Produce in those six months.

We have tried radio ads. They did not work for us. Perhaps in future years, as we become more established, radio ads will become more effective. One strawberry grower in our area said he used radio with great success. But he said that it only started working after he had been in business for many years and a lot of people knew about him.

We rely mostly on display ads in weekly shoppers, local weekly newspapers, and very small display ads in the major metropolitan newspapers in our area. We also have ads in regional tourism publications which work very well for us. We try to keep our ads simple and consistent in appearance.

Soils

Soil acidity (pH) is perhaps the most critical aspect of growing blueberries. Blueberries belong to the botanical family *Ericaceae,* with cranberries, rhododendrons, heaths, and other acid soil dependent plants. This family of plants can grow and prosper on very poor soils as long as there is a minimum amount of iron present, a pH of less than 5.5, and at least 3% to 4% organic matter. Soils with a higher pH will not grow blueberries well, even with large amounts of iron in the soil - it is tied up by the higher pH soils and is not available to the plants for their metabolism.

As soon as soil pH starts to get above 5.5 a characteristic yellowing of the leaves, along with a marked loss of plant vigor, makes itself apparent. The yellowed plant may limp along for a few months, or longer, before it dies. It will eventually succumb to a lack of available iron. There are few soils with inadequate iron if the pH is in the right range. Should you be in an iron deficient soil this problem may be rapidly and inexpensively resolved by adding a few hundred pounds of iron sulfate (ferrous sulfate) per acre banded in the row.

If your soil is very low in organic matter you will have to increase it. Even if you are in the acceptable range for organic material you should plan on extensive mulching to increase organic matter. It improves soil structure, increases moisture supply to the blueberries, and in many other ways provides an improved habitat for the plants. (See the section on organic material for more information.)

In areas with a naturally neutral or alkaline soil (pH 7 or higher) it is not practical to adjust the native soil pH to the appropriate levels. It can be done but two things will happen:

(1) The change in soil chemistry will seriously disrupt the ecology of the soil organisms.

(2) the soil pH will readjust over time and rapidly work it's way back toward the original pH. The finer the soil and the more alkaline the original pH the faster this will occur. You could end up back where you started in a year or two.

This doesn't mean you can't grow blueberries - it only means you have to work harder to be successful at it. It also means that you might not make quite as much money in the long run. But it could mean, if you get production running smoothly and market your product well, really good profits if you have the only high quality fresh blueberries in your area.

How do you grow blueberries in soil that is actually deadly to blueberries? Well, you don't. You need to excavate a trench about two feet deep and three feet to four feet wide and replace the native soil with a manufactured soil. The manufactured soil will be made of approximately:

2 to 4 parts medium clean sand
1 to 3 parts rotted saw dust and/or wood chips
1 part peat moss (optional)
10 to 15 lb. granular sulfur per 100 ft of row
10 lb. of iron sulfate per 1,000 ft of row

These figures are approximate. The pH of your original materials and their response to the acidification process may require adjustments to this mix. You are aiming at a target pH between 4.5 and 5.5 at planting time. The pH of the material at the beginning of the process will likely be between 6.0 and 6.5, and should drop to 4.5 to 5.5 over 4 to 6 months. If pH is not getting close to target two months before planting, till in another 5 to 10 pounds of sulfur per 100 ft. of row.

The sand, rotted saw dust and/or rotted wood chips, and peat moss should be well mixed and placed three to four feet wide and about 3 feet deep in a two foot deep trench. It will settle some but make sure to tamp it well so that after it settles there remains at least a 6 inch mound of bedding material above the surrounding soil. This will look like a slightly raised bed but decomposition of the organic material will level the bed within a few years. Build your planting bed at least several months before your plan to plant, nine or ten months would be better. Build your raised planting beds in early to mid summer in northern states, no later than early fall in southern areas. You will need to spread and mix the iron sulfate and sulfur into the planting bed right after construction and irrigate to keep the bed moist. You need to build up a population of soil bacteria and fungi to keep the blueberries happy.

If you can't find rotted saw dust or wood chips you can use fresh material but you will need to compost it for at least nine months (a year would be better) before you plant. To accomplish this you will need to add about five pounds of ammonium sulfate (21-0-0) fertilizer per cubic yard of manufactured soil mix prior to placing the mix into the trench. If you have to take this route you **must** prepare your planting beds **at least** nine months before you plant - to give the organic material time to break down into a soil like material. You will need to irrigate this material just as if you were growing blueberry plants in it already. The bacteria and fungi that compost the sawdust need water to do the job. If you plant blueberry plants directly into this material before it has composted properly it will probably kill or stunt them - the blueberry plants cannot tolerate concentrated fertilizer in contact with the plants root system. The manufactured soil will also be deficient in available nitrogen, causing lack of plant vigor, until the composting process is complete.

The cost of these materials is highly variable around the country. If you can't find peat at a reasonable price for this project you can get by with a lower proportion of it or none at all. The saw dust or wood chips should be available from saw mills and solid waste facilities for little more than the cost of loading and hauling. A substitute for peat may be approximated by using very well decomposed saw dust or wood chips and adding extra sulfur into the mix at a rate of approximately 1 to 5 pounds per cubic yard. It will take some pH testing here as the normal pH of rotted sawdust can range from 5.0 to 6.8, depending on type of wood, and degree of decomposition. If your rotted wood compost is pH 6.0 to pH 6.5, add 3 to 4 pounds of sulfur per cubic yard. If the pH is closer to pH 5.5, a pound or two of sulfur per cubic yard will be more than adequate. Use a reliable pH meter to test your manufactured soil a few months before you plant so you can make additional corrections as necessary.

Soils with a natural pH of 6.0 to 7.0 can be acidified but will probably require regular addition of elemental sulfur over the life of the planting to maintain an appropriate soil pH level. The frequency of these amendments will be determined by the original soil pH and the soil texture. You will need to do **annual soil tests** at various points around your blueberry patch to monitor pH levels and make necessary adjustments.

The acidification of your blueberry plot in moderately acid types of soils will need to be started nine months to a year prior to planting. The earlier the better. For every point of pH (e.g. from pH 6.5 to pH 5.5) you wish to drop you will need to add, for starters, at least:

Coarse sandy/gravely soil	sulfur =	500lb/acre
Medium sandy loams	sulfur =	2000lb/acre
Fine silty to clay loams	sulfur =	3000lb/acre

These figures are approximate. Soils vary in their capacity to rebound from acidification and in the degree they will acidify. Always monitor the acidification process with a good pH meter. You should monitor the acidification process at least every two months, more often certainly wouldn't hurt. If the soil pH is not approaching 5.5 two months prior to planting additional dispersible sulfur should be tilled in at a rate of 100 to 200 pounds per acre for lighter soils, to 500 to 1,000 pounds per acre for heavier soils.

The per acre rate for sulfur application may be cut proportionally if you only band it in the rows where the blueberries are to be planted. For example if you are planting your blueberries in rows ten feet apart you can mark the rows and apply sulfur 2 1/2 feet out from the center of the row to both sides of the row. You will be applying sulfur to a five foot wide strip every ten feet or over exactly half the field. You have just reduced your starting sulfur requirements by 1/2. (Incidentally, it is difficult to over acidify soils as it will take almost ten times as much sulfur to drop your soil from pH 5.5 to pH 4.5 as it took to get from pH 6.5 to pH 5.5. The pH scale is logarithmic - pH 5.0 is ten times as acidic as pH 6.0., pH 4.0 is 100 times as acidic as pH 6.0, and so on.) Over the life of the planting you will probably need to add sulfur more often to readjust soil pH in the rows because higher pH soil both below and beside the row will tend to raise soil pH in the row more rapidly. Increased soil pH at the row edge may also inhibit root growth at the edges of the rows. A smaller root mass means a smaller, less vigorous blueberry plant.

The other commercial blueberry grower in our area bands the sulfur onto the row prior to planting. Since he is not acidifying the whole field he uses about half the sulfur that we use for initial pH adjustment. Since the rows between the plants are less acidic, by a long shot, I suspect that he needs to side dress a little more often after planting to maintain acidity in the rows. But it seems to work for him very well. I suspect that either method works well enough, and that over the life of the planting total sulfur use will be pretty similar.

We have always just gone ahead and spread sulfur over the whole field at a rate of 2,500 lb. per acre on our silty loam soil (pH 6.5 to 6.7). This has worked well for us. Additional pH adjustments over time are done by side dressing granular sulfur and/or irrigating with dilute sulfuric acid.

Sulfur is available in several forms and can usually be purchased from agricultural feed and chemical suppliers. Regular granular sulfur is widely available, works fairly well, and is clean and easy to handle. However you will need to cultivate the soil several times to allow the granules to break down and mix fully with the soil to get the best pH reduction. Most sources recommend starting sulfur amendments a year before planting to allow acidification to proceed fully.

Powdered sulfur is also available, mixes with soil well, and reacts more rapidly to acidify the soil, but can be very unpleasant and hazardous to use. The sulfur itself is not particularly toxic, but when it mixes with the water in your eyes, mouth, nose, lungs, or sweat it readily forms dilute sulfuric acid - a most unpleasant, and potentially hazardous situation. A full face respirator, protective clothing, and two baths after you are done, are required.

A happy medium, dispersible granules, are available from agricultural feed suppliers and feed mills. This material is actually powdered sulfur compressed into granules. It is easier to handle, but you will want a good dust mask and goggles while working with it, and a shower when you are done. But it breaks up readily with a little moisture and mixes easily into the soil. (Interestingly, granular sulfur is generally considered an "Organic" soil amendment, iron sulfate is not.)

If you really feel that you have to get started planting in a few weeks, or if your initial attempt at pH reduction didn't achieve the desired results you can acidify soil rapidly with sulfuric acid. It will work but may disrupt soil structure and soil ecology. Sulfuric acid has about one half the acidifying capacity of granular sulfur on a pound for pound basis. If your soil tests indicate that you need 2,000 pounds of sulfur, you will need 4,000 pounds of sulfuric acid. At nearly 10 pound per gallon that adds up to about 400 gallons of sulfuric acid. Sulfuric acid **is very, very dangerous_stuff to handle** so there must be an investment in training, proper handling equipment and protective clothing if you want to try this. I wouldn't try this without some serious hazardous materials handling training.

I have used sulfuric acid in irrigation water to help with soil pH control. You can irrigate blueberries with water adjusted to a pH of 3.0 without causing problems with the plants. What will be left of your irrigation system is another question.

If you are using a solid set irrigation system, with aluminum or other metallic pipe and brass or iron fittings and sprinkler heads you will destroy the system if you use slightly diluted sulfuric acid in it. Sulfuric acid eats metal fast. Concentrated sulfuric acid will even degrade many plastics. For more information about irrigating with pH adjusted water see the section on irrigation.

Always check your soil pH a few months prior to planting to see how the pH adjustment process is coming along. You should aim for a pH of 4.5 to 5.5 prior to planting. If, a month prior to planting, the soil pH is getting near your target range, check to see if you can still locate sulfur granules in the soil. If there is still some sulfur visible, little yellow specks and lumps, your pH will continue to drop for a few months. You can go ahead and plant. If you are still a half a point away from your pH target and cannot locate or identify any residual sulfur in the soil you should add more. Apply about 5 pounds of dispersible sulfur per 100 feet of row banded in the rows.

Once your pH is correctly adjusted (have this verified by a soil testing laboratory) and your blueberries are in the ground and have started growing you will still need to monitor your soil pH on a regular basis, at least once a year. A good pH meter is a must if you are serious about growing blueberries. You will have to spend a minimum of $50 to get a real pH meter - $150 to $300 to get a really good one. Any pH meter that is worth purchasing will come with small jars of calibration fluid so that the meter can be adjusted and checked each time before it is used. Absolute accuracy is less important than knowing the meter is operating properly. Accuracy to pH 0.2 units is OK. The pH meters sold in most gardening magazines for $15 or $20 are junk.

Using a consistent procedure is important when you are monitoring your soil pH. Soil samples should include at least a third of your mulch material and should be taken to a depth of six to nine inches. (Remember you should calibrate the meter first. Consult the meter's operation manual for calibration procedures.)

In the usual laboratory procedure you (1) take ten grams of soil and ten grams of deionized water, (2) mix them, by shaking, for a minute or two,

(3) let the solution settle for five minutes, (4) insert the meter into the clear portion of the mix above the mud and, (5) take the reading off the meter. Several soil samples should be taken from around your blueberry planting to assure a representative set of readings. Each sample should be tested separately to see if any particular areas are changing pH more rapidly than others and to check the consistency of application of sulfur. You will probably find some variation in soil pH due to changes in soil types within your blueberry patch. Local areas that are becoming less acid should have extra sulfur applied if the pH is nearing 5.5. In most cases you don't need to worry about getting too acid.

On some types of soils, however, notably gravelly silts and sandy loams, it is possible to have a pH "crash" after several years of soil acidification. This occurs when enough sulfur or ammonium sulfate has been added to neutralize the limestone or carbonates naturally existing in the soil, and the soil looses it's buffering capacity. This is another good reason to monitor your soil pH regularly. In some cases pH could drop to levels too acid (pH less than 4.0) for good blueberry growth and production.

As I mentioned earlier, our soil is naturally about pH 6.5 to 6.8 and very fine textured. These are not the best parameters for starting a blueberry patch but we went ahead anyway. Our first field was worked over with a plow and disk in early summer and we ordered 2000 pounds of granular sulfur from a local supplier. Well, we did have to explain ourselves a little because nobody around here uses sulfur in that kind of quantity. But a local agricultural chemical supplier delivered the sulfur in a big spreader which I hooked up to our small 1941 Farmall A. Of course we are on a hill and it had rained recently, so I promptly got stuck and had to run up to the neighbors for more power.

I planned to disk the sulfur in the next day. But we had a big time thunderstorm that night. I couldn't disk, some of the sulfur had washed off the field, and boy, did that field stink. It could make your eyes water and your nose hurt just looking at it. That is caused by vaporizing sulfuric acid. It is like getting a brief whiff of the brimstone part of fire and brimstone. It finally dried out about a week later and we got the sulfur disked in - and the field still smelled like sulfuric acid. I had expected (hoped?) the acidification might drastically effect the growth of the usual weeds in the field but they came up just as strong and healthy as ever.

We followed this same procedure almost every year for the next seven years on each successive addition to our blueberry patch. Some years we would acidify two acres, sometimes only one. We had soil tests done the first year and they indicated that we had dropped our soil pH to exactly pH 5.5. (It was so perfect that I still think somebody got something wrong.) We went without soil tests for a few years after that - a lapse of judgment I don't recommend.

My solution to the ongoing adjustment of soil acidity is to use ammonium sulfate fertilizer for a nitrogen supplement (ammonium sulfate - (21-0-0) is about 20% sulfur) and to have my fertilizer supplier add an extra 10% to15% sulfur into the ammonium sulfate.

In an emergency, i.e. when we spotted a sick, yellow blueberry plant, I add some iron sulfate around the roots of the affected plant. For the half high varieties we sprinkle about a quarter to a half cup of this material, a cup or slightly more on highbush plants. We spread the iron sulfate around one to two feet out from the center of the plant. Iron sulfate is an acidic granular form of iron which can make a remarkable difference for a down and out, yellowing blueberry plant. It is somewhat water soluble so an immediate watering after application brings fast results. I have heard of dissolved iron sulfate being used as a foliar feed (sprayed on the leaves of the plants) in areas with higher pH soils as an iron supplement. It may work, I don't know.

Many gardening supply catalogs offer Aluminum Sulfate as a soil acidifier. It works well for small applications near azaleas, hydrangeas, and even blueberries. It is too expensive for major pH adjustment on a commercial scale. It has only 1/6 the acidification potential of granular sulfur. Also, in some soils, use of too much Aluminum Sulfate can cause aluminum toxicity in blueberry plants.

Organic Matter

Another important aspect of soil makeup to consider is organic content. Blueberries require an organic content of at least 3%, and 20% to 30% is not excessive. If you are starting out at the lower end of the scale you will need to develop a long term program to increase your soil's organic content. This will include starting out with a green manure cover crop prior to planting, possible spreading and working in of peat or rotted saw dust, tree bark, or wood chips into the soil prior to planting, and an extensive and regular mulching program after planting.

A high organic content soil provides good soil structure and permeability, aids in water infiltration and retention, provides a good habitat for beneficial soil organisms, and moderates soil temperatures. There is probably nothing you can do to benefit your blueberry plants more, once soil pH is right, than give them lots of mulch. A mulching program adding 3 to 6 inches of mulch in a band 4 to 6 feet wide on each row every two or three years is a good target, but represents a great amount of labor. Nitrogen application should be increased by about 50 lb. per acre if fresh, non-composted saw dust or wood chip mulch is applied.

Blueberries need a constant supply of water and really like a loose friable soil to spread their roots around in. Normally they grow in sandier soils which allows for good root growth, but unless there is adequate water, crops are irregular and berries are small. Most gardening magazines will recommend adding peat to the soil for growing blueberries and this is an excellent idea. For the larger commercial planting peat will get expensive. I have used both oats and buckwheat as a green manure crop to increase soil organic material with some success.

Green manure refers to plants grown for the sole purpose of working them into the soil to add organic material to the soil. Buckwheat is a popular green manure crop that tolerates acid soils well, and is ready to plow down in 5 to 6 weeks (before seed is matured). Oats will work well if planted early in the spring or early in the fall. Even weeds do a good job of providing organic material - just be sure to plow them down before seed maturation.

Blueberries love mulch. I strongly recommend that you put down as much mulch as you can get your hands on. We chop bales of hay onto our blueberry rows, leaving a 2 to 3 inch mulch that decomposes within three or four months. I don't feel that this is enough. I would like to see at least four to six inches put on every year. Mulch serves to hold moisture in the ground and keeps the soil cool. It adds organic material and nutrients to the soil continually. It also serves as a soil conditioner adding organic material and feeding billions of soil bacteria, fungi, and larger critters that make up a healthy soil. Wood chips, saw dust, chopped hay or straw - use your imagination but keep adding mulch to your blueberries. Your plants will thank you and your customers will appreciate not standing on dirt.

Several studies indicate that saw dust, either hardwood or softwood, works best for mulching blueberries. Tree leaves and straw/hay were a substantial improvement over no mulch, but not as effective as saw dust. Any mulch is far better than no mulch, unless it has lots of weed seeds and you aren't using any herbicide.

Pocket gophers love mulch also. It makes a great habitat for them. (See the section on pests for more information.)

Eventually you may want to get your own chipper/shredder to make mulch with. I have spent a lot of time studying what's available. I haven't got a real one yet but when I do I know what I'll look for. Plenty of power (20 to 40hp), you will want something that chips at least 6 inch material, and this means a PTO driven tractor mounted unit. Also, you will want adjustable or multiple screens to vary your chip size. Optimally you will want to make a mixture of 70% to 80% finely chipped material, almost fluff or fine enough that it won't give away free splinters, and 20% to 30% coarser chips the size of a matchbook. The coarser chips allow water to penetrate the finer mulch more easily and block weed development better. The finer material holds moisture better, breaks down a little faster, and is more comfortable to sit or stand on.

Lacking a real chipper/shredder, as usual, I make do with the wrong piece of equipment for the job. We bought a bedding chopper (bale chopper) to mulch the blueberry plants with. It has an 11 hp engine and is designed to chop bales of hay or newspaper into bedding for dairy cows. We have bolted it to a small wagon and Terry drives the tractor s-l-o-w-l-y down the blueberry rows as I feed bales into the chopper. It works all right but there must be better solutions to this problem. The bale chopper also handles blueberry prunings and small apple tree prunings, anything up to about an inch and a half. This constitutes rough service for this type of equipment. Don't expect much warranty service from the manufacturer if you tell them you have been chopping up branches in their bale chopper.

If your are making your own mulch and putting fresh, uncomposted mulch on your blueberry plants, you will have to increase the amount of nitrogen you apply as fertilizer to feed the bacteria and fungi that are breaking down the wood as well as the blueberry plants. If you don't feed them extra nitrogen, about 10 lb. actual Nitrogen per acre for each inch of fresh chipped mulch you apply, the bacteria and fungi decomposing the mulch will steal nitrogen from the blueberry plants and reduce their growth.

If you have the time and the material you can compost your mulch prior to application. It will take a year if you add Nitrogen fertilizer at the rate of 1 to 3 lb. of actual N per cubic yard of chips. You can compost the chips over a longer period by using less nitrogen.

After a few years of adding nitrogen and mulch, or composted mulch, you may be able to reduce nitrogen inputs to the field by a moderate amount as there will be nitrogen made available to the blueberry plants from the decomposing mulch. Don't expect to eliminate N inputs altogether because you will still need to be adding fresh mulch every year or two and this will, unless already composted, require additional nitrogen.

Planting

Take good care of your blueberry plants after they arrive from the nursery. Make sure that you order them to arrive at a time when you can get them in the ground in a timely fashion. Your blueberry plant supplier will cooperate to the best of his ability on getting your plants to you on your schedule so they can be planted as soon as possible. Stay in contact with your supplier as delivery and planting time approaches to avoid mixups. In the best of all possible worlds the blueberry plants should be planted back in the ground 3 or 4 days after being shipped from the nursery. They do not tolerate drying roots at all.

As soon as the plants arrive, they should be placed in the shade and watered. To minimize handling, they may be placed on plastic laid over a wagon parked in the shade. They should be watered and covered loosely with plastic to minimize drying out. Be sure to open them up during the day to let them breathe. As soon as the field is prepared enough to start planting, and the planting crew is ready, the wagons should be rolled out to the field and planting commenced.

Getting the plants in the ground fast is a problem we were lucky enough to solve early in the process. There are two steps: (1) prepare your ground. The field should be acidified the year before planting, sprayed to remove perennial weeds, plowed and disked until loose to a depth of 4 to 6 inches, and with rows staked out and string lines set between stakes. Shovels, measuring sticks, and irrigation supplies should be set out at the ready. (2) Have a Party - a Planting Party and invite all your friends.

Planting the new blueberry arrivals is probably the most time critical, and most labor consuming operation you will face in starting and operating a berry growing operation of any sort. It takes 5 to 6 minutes to plant a two or three year old blueberry bush properly in well prepared soil. If you want to plant 1,000 plants it will take you 100 hours. If you divide that time up between two or three people it will take them a week to plant all the blueberry plants. The last plants put in the ground will be badly stressed and the people involved will be exhausted at the end of the week. If you divide the work up between 20 to 30 people you can get the whole operation done in 3 to 5 hours. Everyone involved will have a great sense of accomplishment and enough energy left over for a picnic.

You might think that this is taking advantage of friends. Our experience is that it is very easy to get people involved, that they look forward to coming back next year, and are disappointed if there isn't a reasonable amount of work to be done. Your friends will actually recruit more participants for you. I must add that, as an additional incentive, Planting Party participants get free berries for life. This is a really cheap way of saying thank you.

We started in 1987 with about 18 people at the first Planting Party. We planted about 1,000 half-high blueberry plants in 5 hours. It was hard work because nobody, including us, knew what we were doing. Over the years participants made suggestions, we learned more about soil preparation, more people were there to help, so the planting got easier. In 1994 we planted 1,200 highbush plants in just over three hours and nobody, except a few hole diggers, even broke into a sweat. The planting went so fast that, near the end, 300 foot long rows of blueberries were going in faster that I could hook up the irrigation lines for them.

Of course you can hire temporary labor to do this job but they will not work as efficiently as a group of friends. The project will be a chore rather than an interesting morning or afternoon's work. And there is the cost and uncertainty of trying to round up 30 people willing to work for a few hours on a particular day. I really recommend to the Planting Party approach to this part of the project. After all, If you can't get your friends interested in your business, how will you get strangers interested?

Fertilizer

In the wild the blueberry plant tends to grow on soils low in available nutrients, pH 4.0 to 5.5, and usually with an organic content over 3% to 4%. I have seen wild blueberries growing in soil so poor that they had very little weed competition of any sort, just a few jack pine trees, blueberry plants, and bare sand. In commercial production, the blueberry plant will benefit greatly from better soils, some fertilization, and irrigation. The increased vigor and production will come at a cost of increased weed competition.

On most soils of even marginal fertility the only fertilizer input regularly required for commercial production of blueberries will be nitrogen. Commercially available nitrogen fertilizer comes in several forms. Ammonium Sulfate is a readily available granular form of nitrogen fertilizer. It contains about 20% nitrogen by weight. If you want to add 50 lb. of nitrogen to an acre of blueberries you will need to spread about 250 lb. of ammonium sulfate. Ammonium sulfate is the nitrogen source of choice for those of us trying to grow blueberries on soils needing acidification. The "ammonium" supplies nitrogen to the plants and the "sulfate" supplies acidity to the soil.

Urea is another form of nitrogen fertilizer. In its granular form it is a more concentrated source of nitrogen (about 46%) and also acts to acidify the soil, but to a much lesser degree than ammonium sulfate. The fact that urea is more concentrated can easily cause problems if you are spreading it on the rows. A little spill, or uneven application, can over fertilize plants. You can literally burn the blueberry plant to death with excess nitrogen fertilizer.

There are spreaders manufactured to side dress fertilizer onto row crops. They cost from $800 to $1,000. I want one but I don't have one yet. For the last eight years I have been using a $30 drop spreader designed for fertilizing lawns. I never installed the handles, I just wired the thing onto the rear of my tractor seat, put a piece of rebar through the wheel that rotates the distribution paddles. I wiggle the rebar, and drive down the rows dropping fertilizer over the top of the plants. It leaves me with a sore arm but I can fertilize seven acres in a day. Of course I am going to have to upgrade to a better fertilizer spreader soon. My old Farmall A won't go over the top of a 5 foot tall high bush blueberry plant.

I have the fertilizer dealer mix 10% to 15% extra granular sulfur into the ammonium sulfate for more acidification. I need to add extra sulfur every few years to keep the soil pH down. With the sulfur added to the ammonium sulfate fertilizer it just takes one trip over the fields instead of two.

Blueberries like iron but they can't get it if the soil pH (acidity) is over 5.5. Typically the bushes will lose vigor and the leaves will turn yellow. Plants with "Iron chlorosis" respond well to treatment with iron sulfate. Iron sulfate is an acidic form of iron that is readily available to blueberry plants. Shaking a half a cup around any yellow half high blueberry bushes, more around highbush plants, should perk them up in a week or three. Some extra granular sulfur at the same time will usually effect a longer term cure.

Phosphorous (P) and potassium (K), two other important plant nutrients, are usually available in adequate amounts to grow healthy blueberries in most soils. Commercial blueberry growers who have older plantings on poor sandy soils may benefit from additional P and K in their fertilization program. In some areas some soils may lack other trace nutrients. The book "**Highbush Blueberry Production Guide"** (see Appendix A) has an excellent section on Blueberry nutrition and trace mineral requirements. I recommend it to those few growers who may develop problems in this area.

Stay away from any fertilizers containing chlorides or salts. Blueberries do not tolerate salt (sodium chloride) or any chemicals with chlorides in them.

Manure, either composted or fresh, can be used on blueberries with excellent results. Since I don't have a ready supply I use commercial grade ammonium sulfate at a rate of 50 pounds actual nitrogen per acre each year. Using manure will probably be more work in term of spreading it in the rows but the blueberries will appreciate it just as much as commercial fertilizer. Application rates for manure vary with the type of manure. Poultry manures usually contain higher amounts of nitrogen than cow or horse manure. Fresh manure is "hotter" than aged or composted manure.

Fertilizer at planting time will damage the roots of the blueberry plant, often to the point of stunting or killing the young plants. The only situation which may require the addition of fertilizer prior to planting would be the use of fresh sawdust or wood chips as a soil conditioner. This process must be started 6 to 9 months prior to planting to allow the fresh sawdust to break down and the fertilizer to be processed by soil organisms.

Light applications of nitrogen fertilizer may be started about a month after planting. About 5 to 10 pounds of nitrogen per acre may be applied at 4 and 8 weeks after planting if desired. It is preferable to apply this nitrogen directly to the soil, in a 1 to 2 foot circle around the base of the plant. Application should be by hand or by chemigation. Don't apply over the top of the plant because the nitrogen can burn fresh young foliage. I don't think these early applications of nitrogen are absolutely necessary and I have never used them. Your time and effort is better spent on good soil preparation prior to planting, and weed control after planting.

Chemigation is the application of fertilizer, or other agricultural chemicals, via irrigation water. In blueberry culture there are several areas where chemigation may prove helpful. I think the largest use of chemigation in blueberries is injection of acids into the irrigation water to control irrigation water pH. Optimally your irrigation water will have a pH of around 5. If your water is that acidic you are lucky, our well water is about pH 7.3.

Irrigation water that alkaline will cause blueberry roots to grow away from the irrigation water. I have seen pictures of very lopsided root systems, caused by alkaline irrigation water, where the side of the plant away from the drip irrigation line had bushy, vigorous roots, and the side toward the irrigation line had short, weak roots. Irrigation water pH of 6 or above may cause problems over the long term. Water pH above 7.0 may cause problems in a year or less. Rather than adjusting water pH each time I irrigate I usually adjust the water pH to around pH 3.0 for one irrigation using sulfuric acid. This "stores" some acidity in the soil that will, over time, be neutralized by the more alkaline irrigation water applied later.

On lighter sandy soils it may be beneficial to chemigate some of the required nitrogen onto the field. Apply half the required nitrogen in granular form prior to leaf out in the spring, and the other half may be applied in the irrigation water in May and early June. Both granular urea and ammonium sulfate are readily soluble in water. This is actually a problem as it requires someone to be available to continually add small amounts of the granular fertilizer as the irrigation water is applied to the field. Luckily Urea is also available as a liquid (usually 28% Nitrogen) so fertilizer injection can be automated. An alternative is dissolving ammonium sulfate or granular urea in a drum of water and injecting the solution into the irrigation system. I haven't done this but it should work out well enough.

I also use my chemigation system to apply iron in the form of iron sulfate. Iron sulfate is not highly soluble in water so this works out to be pretty easy. I just toss a 50 lb. bag of iron sulfate into my mixing tank and continue to run the irrigation water through it. It takes several days for the Iron sulfate to dissolve completely. I only need about two bags and I can give seven acres of blueberries a mild shot of acidic iron as a booster.

On heavier soils N may be applied in one application prior to leaf out. In lighter soils use a split application, one half prior to leaf out, one half a few weeks after blossom drop, in late May or early June. Since my patch is on a heavier soil I apply my nitrogen/sulfur in a single application just prior to leaf out in late April or very early May. The heavier soil prevents excessive leaching of the nitrogen and the blueberry plants get off to a good start in the spring.

I can think of no situation where any nitrogen fertilizer should be applied to blueberry plants after the middle of June. Late applications of nitrogen will cause a late flush of new cane growth that will not harden properly for winter. Most of this growth will winter kill. It may also force plant energy into vegetative growth in the late summer, during the period when the plant should be making flower buds for next year's crop. Even applications of nitrogen in June, just prior to ripening, are reputed to affect berry flavor and shelf life. If you are doing split applications of nitrogen, plan on doing your last application a minimum of 3 to 4 weeks prior to ripening time.

Irrigation

After proper soil acidity, and a good mulch, adequate water supply is the most important factor in blueberry production. It may seem like rainfall is adequate but blueberries require at least an inch to two inches of water a week until they are done fruiting, and they still need at least half that amount of water after they fruit because they are making fruit buds for the next season. You may think that it rains enough in your area - but it hasn't, it doesn't, and it won't. In the upper midwest the weather is nearly always dry for 3 or 4 weeks in late April and May, right around planting time. It is almost inevitable that there will be another week or two between then and harvest that the rain takes a week or two off. If you don't irrigate as needed your berries will not get as big, your plants will get stressed, and you will probably affect flower development for the next year's crop.

If you have ready access to a lot of water (100 gallons per minute or more) you can put in any type of irrigation you want. An overhead sprinkler system is a good choice because not only can you irrigate the berries, the irrigation system can be used for frost protection. Of course the electricity costs will be generally higher than with a lower flow system. You also should check in with your state Department of Natural Resources about regulations involving irrigation from wells and/or ponds, and to get the necessary permits.

For those of us with deep wells and lower water production capabilities drip irrigation is the only reliable (sort of) method of water delivery. We have a 430 ft well, with a 3 hp pump that can deliver about thirty gallons of water a minute. With drip irrigation I am able to irrigate about 2/3 of an acre in 10 to 12 hours if it has been really dry - 8 hours for a lighter watering. With seven acres of blueberries, and two and one half acres of raspberries, I can end up running the well for five or six days straight when the weather gets dry. Some years that is five or six days straight, week after week after week after week. My record so far is 15 weeks of irrigation in one season (it really wasn't that dry, just very poorly timed).

With our well and pump it costs about $25 per complete round of irrigation for the power to pump the water. Since we are very close, or maybe a little over the pumping capacity of our well, we face a choice. No further expansion of crops requiring irrigation, or build a pond for additional irrigation water, or drill another well.

Another problem with our irrigation system is water quality. It is great drinking water because it comes from 430 feet down in a limestone aquifer. But because of the limestone the pH of the water is about 7.3 and water that alkaline will cause blueberry roots to grow away from the irrigation water. If soil acidity has been adequately controlled this is not an immediate problem, but over the long term high pH irrigation water will raise the pH of the soil out of the range that is healthy for blueberries. There are two solutions to this problem, and I use both.

First I have a regular program of soil acidification. I use ammonium sulfate as a nitrogen source. Each pound is about equivalent to 1/6 of a pound of elemental sulfur in it's ability to acidify the soil. I add about 15% additional elemental sulfur to the ammonium sulfate which makes it twice as acidic. Every few years I add about 5% (by weight) iron sulfate to my fertilizer mix for additional iron and additional acidity. Of course I check soil pH regularly and adjust fertilizer components accordingly.

Another method of dealing with high pH irrigation water is to acidify it prior to putting it on the field. This is usually accomplished by injecting concentrated sulfuric (or phosphoric) acid into the water in the irrigation system. When I irrigate with sulfuric acid to adjust the water pH, I continually test the water to keep it's pH around 3.0, about the acidity of lemonade. Since I use plastic hose and plastic drip irrigation line this level of acidity is not a problem. I do use a few galvanized pieces of pipe in the system but it will take a few weeks to cause enough damage to affect the pipe.

When adjusting irrigation water pH, I pump water into a tank (250 gallon polyethylene) raised on a wagon for additional head. The whole system is now gravity feed to the irrigation system. (With less water pressure I have to irrigate each area for a longer period.) I do live on a hill and all the blueberries are down hill so the gravity feed system will work. For flat land situations a more sophisticated injection system will be needed.

Using an adjustable rate chemical feed pump and some very acid resistant tubing, I drip concentrated sulfuric acid into the raised tank and let the inflow of water mix the acid. **Always operate chemigation systems with a siphon break between the water supply and the injection point. It is imperative that, if your water pump should fail, no chemical/water mix can possibly get siphoned back into your well or water supply. Check with your County Extension Agent for local regulations covering chemigation, and for proper safeguards to protect your water supply.**

Test the outflowing water several times for pH level right at the start and adjust the feed pump flow accordingly to maintain irrigation water pH at or near 3.0. Continue checking pH at intervals of several hours to ensure a fairly stable pH between 2.7 and 3.3. This system has worked well so far. **Again, concentrated acids are hazardous. It will eat holes in you, your clothes, and just about anything it touches. If you try this do your research and get hazardous materials handling training. <u>I do not recommend this procedure to anyone not trained in handling hazardous materials.</u>**

Timing is very important when irrigating your crop. You don't want to deprive your blueberry plants of water when they are making berries or the crop will be reduced. Blueberries are about 95% water. Even water stress after the harvest can affect the formation of flower buds for next year's crop.

On the other hand you don't want to over irrigate because it costs money and time, and you don't want to waterlog the soil. This can injure the berry plants as well. Blueberries need 1 to 2 inches of water a week while they are making berries, or from leaf out until the crop is picked. After that they should have at least one half to one inch a week, on average, until they go dormant in the fall. The higher rates apply to light sandy soils in hot weather, the lower rates apply to heavier soils with an organic content of 5% or better.

There are several devices that will measure soil moisture for accurate prediction of irrigation needs. A tensiometer consists of small water filled porcelain or plastic cups which are buried in the soil and connected to vacuum gauges by tubing. A change in soil moisture causes a change in vacuum reading.

Another device is the electrical resistance block. A block of gypsum is attached to two wires and buried at an appropriate depth in the soil. Soil moisture is indicated by measuring the electrical resistance of the gypsum block, which has less electrical resistance as the soil gets wetter. In either case the sensor for these devices should be buried at a depth of 6 to 12 inches. If you are using a lot of mulch, in excess of three inches, your sensors should be 6 to 8 inches deep. With less mulch or in very light sandy soils you sensors should be 8 to 12 inches down.

Another method of irrigation management considers the water holding capacity of the soil and the normal water consumption of the plant. The blueberry plant can consume up to 0.25 inches of water per day while growing rapidly and making berries during hot weather.

The blueberry plants roots occupy about the top foot and a half of the soil. Different soils have different water holding capacities. Sand and gravel soils hold very little water, perhaps 0.7 to 0.9 inches of water per 1.5 feet of soil. A sandy loam soil may hold 1.9 to 2.25 inches of water per 1.5 feet. A silty clay can hold 3.75 inches of water per 1.5 feet of soil. (Addition of significant organic material will increase water holding capacity of all soils, but will benefit lighter sandy soils most.)

You do not want to reduce the water in the soil below 50% of capacity or you will begin to stress the blueberry plants. If you have a light sandy soil with only 1 inch of water stored in the blueberry root zone you will need to replace that 0.25 inch of water per day the plants use **every other day** during the peak growing season. If you are growing blueberries on a silt loam that stores 3 inches of water in the root zone you will only need to irrigate every six days but you will need to replace an inch and a half of water at that time.

Of course I use a very sophisticated system which coordinates the use of a rain gauge and a finger stuck in the dirt. If it hasn't rained an inch in the last week and the dirt feels warm or dry I irrigate my berries. If it has rained, but not enough, I may wait a day or two and then start to irrigate, or I may start right away and just irrigate for a shorter period of time. If there is a good chance of substantial rain predicted I usually wait. If the weatherman says nothing is coming I'll start right away.

The organic content of your soil, the amount of mulch you use, and the type of soil you have will strongly affect the amount and timing of irrigation. On the dry end, e.g. soils of low organic content, with clean cultivation, and no mulch on a light sandy soil will require a larger overall amount of water per week, and it will need to be applied more often. Instead of one 12 hour irrigation per week, perhaps three or four 5 hour periods of irrigation will be necessary. On the other end of the scale with a loamy, high organic soil, and a heavy mulch, the irrigation schedule may be stretched out to one or two inches applied every 8 or 9 days because of higher water holding capacity and cooler soil temperatures. This is a great argument for mulching your blueberry plants very well.

An acre/inch of water is 27,154 gallons of water. If you need to put down an inch of water on an acre of blueberries that is what you need to deliver. If you are irrigating with an overhead system, so you are covering the whole acre, and you are pumping 100 gallons per minute, you will have to irrigate for 4.5 hours to deliver that much water. (4.5 hr. x 60min. x 100gal. = 27,000 gallons) This ignores the fact that overhead irrigation is less than 100% efficient as water evaporates and blows out of the target area. To be on the safe side you should irrigate for 5.5 to 6 hours. Always spend time in the field to monitor the results of your irrigation. It's the only way to know what you are really accomplishing.

Calculating irrigation needs for drip systems is different than with overhead irrigation. Since the drip tape is right next to the berry row, and water is applied at a low rate, there is little or no loss to either evaporation or watering areas away from the plants. With our particular planting configuration I estimate that I am applying water to half the actual area of the field. If I was irrigating an acre of blueberries I would need to apply about 13,500 gallons (1/2 of 27,000) of water. Actually I only irrigate about 2/3 of an acre at a time, because that is all I can keep pressurized with my pumping system, so I only need to deliver about 9,000 gallons of water each time I irrigate a particular section of field. At 25 to 30 gallons per minute, my pumping capacity, I can deliver this amount of water in about six hours. I end up irrigating each section for about 8 hours to allow for variations in water pressure in the system. I'd rather irrigate a little extra than not enough.

Drip irrigation is more efficient than overhead with respect to water use, but it has it's own set of problems. Thirsty little critters with sharp teeth will perforate extra drip holes all over the place. The entire system will need to be checked for larger leaks every week. The best way to check for leaks it to let the system pressurize for an hour or two and then feel the far end of each drip tube for pressure.

Significant leaks will show up as flat or low pressure tubes. They may also show up as very fine six foot high fountains of spray. Leaks which are not visible are usually audible as high pressure water hitting mud. When all else fails look for large wet places. You will learn quickly how to patch the drip tape.

I have tried several different weights and grades of drip tape and have had the same experience with all of them. The heavier grades last a little longer than the lighter grades but they end up with patches all over the place and need to be replaced in a year or two. I finally got sick of searching for leaks and patching in the spring. Now we use the lightest (4 mil.) drip tape and replace it every year.

Pests

For the full rundown on every conceivable insect, fungus and disease I'll recommend **Blueberry Culture**, by Childers, and/or **Highbush Blueberry Production Guide**, by Northeast Regional Agricultural Engineering Service.

If you are starting a blueberry patch and you don't have other large commercial patches nearby you will probably find you actually don't have much disease or insect pressure. Our patch is over 50 miles from any significant commercial blueberry planting. While I have seen insects in the field and probably have some minor problems I have only needed to spray for insects once. This year, on two fields of very young non-bearing blueberry plants, an epidemic of "spit bugs", a type of aphid, appeared on about 50% of the growing tips of all plants. There was a lot of wilting and poor growth. I had to actually wash each infestation off with a light insecticide solution and a back pack sprayer.

I also used a fungicide once. We had a bad hail storm and some plants got beat up and bruised. Some people from the University of Minnesota said that there would likely not be any problem, but to be on the safe side I might want to spray them to protect against fungal infection. I sprayed and had no problems, probably wouldn't have had any if I hadn't sprayed.

You will also have to deal with certain pests that like to eat your plants or berries.

Probably the biggest pest problem you have will be birds. Cedar waxwings and robins can, and will, live on nothing but ripe or nearly ripe blueberries for as long as they are available each summer. They have a complex communication system which will alert every bird within three days flying distance that **it is worth the trip.** Your robin population will increase every year. The cedar waxwings will find your patch after a year or two and will arrive like clockwork every year after that in flocks of 20 to 100 birds. They bring an appetite and a lot of friends. There will be plenty of other species of birds which stop by for a snack. What they eat is bad enough, but once the blueberries are really ripe they will knock 2 or 3 berries to the ground for every one they eat.

When I was a kid in New Jersey my grandmother planted four blueberry bushes in the back yard. They made beautiful flowers, and we would watch the berries develop for several weeks. As soon as they started to turn from dark green to white (about 10 days before ripening) they were gone. I don't think I ever got more than a handful of tart, half-ripe berries. The robins and catbirds just made them disappear. After a few years we covered the plants with cheese cloth. We got a few more berries after that, but we also caught a lot of birds in the netting.

Left uncontrolled on a larger planting, you will loose hundreds of pounds of blueberries every day to the birds. A hungry, healthy flock of cedar waxwings could clean off a small patch of blueberries in a few weeks, and they will have plenty of help. On larger plantings birds will eat 20% to 50% of your crop, and your profits. The only 100% (almost) effective control is netting. If you are planting an acre or less you should give netting serious consideration. I have heard of a blueberry patch, about 60 miles from here, that has netting over two acres. Obviously this costs money, on the order of $1,000 to $2,000 per acre. The netting itself will cost a few cents per square foot and the support structure will not be cheap. There will also be labor costs involved in installing and removing the netting each season and in repairing wind damaged netting. While balancing these costs against actual losses to birds leads to several "guestimates", I suspect, from my experience, that the reduction in losses to the birds will pay the cost of the netting and structure in the first year.

I have felt for several years that I would never use bird netting but last year I changed my mind several times, in both directions. I noticed that the robins are out there in the patch weeks before the berries are really ripe picking them as they start to turn from green to light yellow. While we have been managing a 25 day picking season for our customers the robins and cedar waxwings are picking berries for 40 days, at a rate that may exceed 100 pounds a day! So we have been pricing netting and poles to cover seven acres of blueberries. We can't afford to do it all in one year but we planned to do about two acres next year. but, if we don't do it all in one year the birds will simply move from the covered patch to a field not netted yet and consume berries there at an even faster rate.

Prior to putting in the netting we talked to several growers who use it. Their strong recommendation was that if we were in a windy area to forget it. The wind would ruin the netting. Sure enough, two weeks before the berry season started a 60 mph wind storm took limbs off our trees, shingles off our buildings, and would have delivered our netting into the next county. We has changed our minds again and ordered three electronic recorded bird distress call "squawk boxes".

These squawk boxes are a newer technology in audio bird repellents. These are basically cassette tapes or electronic recordings of bird distress cries, (can I volunteer to get those birds to give out distress calls?) For the small grower they are pretty pricey ($150 - $500) but they are effective for a period of time. This year we used three units to protect about 5 acres of berries. These units are advertised to cover up to two acres each. They run off of lawn mower batteries for a week to ten days. They proved to be very effective at restraining bird activity for three to four weeks in our berry patch this year. The units were mounted on poles a few feet above the height of the berry plants. Volume and frequency can be adjusted and should be varied every few days. I also placed some realistic plastic owls at the top of the poles for an added scare tactic. An occasional shotgun blast across the field served both as a further deterrent and as a means to gauge the effectiveness of the squawk boxes.

In prior years a shotgun blast across the field would scare up 20 to 40 hungry birds but this year usually raised only 5 to 10. That translates to an 80% reduction in bird losses. But the birds will figure it out in a few weeks and your losses will increase again. Remember that these devices, or any bird deterrent activity, must be started at least two weeks before the blueberries start to ripen - once the birds get used to eating in a field it is hard to change their minds.

Propane cannons are the most common method of bird control used by commercial growers. They make a **very** loud bang at preset or random intervals, the birds will avoid the area for a week or more - so will everyone else. You should notify your county sheriff and you neighbors if you plan to use propane cannons. They probably won't work if you sell U-pick. The noise will likely drive your customers away as well as the birds. Propane cannons are pretty effective at scaring birds out of the berry patch, but only for a week, maybe two weeks at the outside. Then the birds will get used to them and you will have to try another approach.

There are also visual repellents like plastic owls, scare eye balloons, and flashy mylar tape. In 1995 I tried Scare Eye balloons and black plastic "Hawk Shadows" that I cut out of heavy black plastic and set out in the field. They seemed to keep the cedar waxwings away for about a week, but not the robins. In that first week I expect they paid for themselves several times over as cedar waxwings can eat a lot of berries. They must be moved often or the birds get used to them and will loose their fear of these deterrents.

There are several wild ideas which you might want to try if they feel right to you. Training a bird dog to actually run the rows and chase birds should keep birds out but it may leave you with an exhausted dog. Installing hawk perches around your fields may encourage some hawks to hangout in the area and actually eat some of the birds. A variation on the hawk theme would be to use a remote controlled airplane to chase the birds off. I recently got a suggestion that I contact a local falconry club and allow them to exercise their birds at my blueberry patch. This is a hot idea that is worth some follow-up. I've seen red tail hawks clear a field of cedar waxwings with just one pass. If you like wild ideas try these out and let me know how they work. You will find yourself thinking a lot about birds if you grow blueberries.

Besides birds you will probably have some damage from deer but it may not be significant. We see deer, and sometimes wild turkey, in the field at the same time as people are picking berries. If you are selling berries U-pick this is a selling point, not a problem. If it becomes a problem you will have to fence the patch, which is expensive, or you can hang cakes of soap around the patch, which is cheaper but not 100% effective. Other alternative are available. Check with you County Extension Agent for other ideas.

If you have bears in the area you may well have these larger visitors. I have not had the problem even though there is supposed to be a family of black bears living in our valley. The reason they don't stop by our place, I believe, is that we are a U-pick and there are always pickers in the field during the day, and the patch probably smells like humans all night too.

Pocket gophers will be problematic if they live in your area. Not only do they like to eat blueberry roots and live in undisturbed soil, they really like to live under a mulch. It's nice insulation for their burrows and sleeping areas. They like to eat blueberry roots in the early summer. If that isn't bad enough, dogs, coyotes, and badgers like to try to dig them out of their burrows. That makes a big hole, often right where a blueberry plant used to be. At best you have a big hole to fill in; at worst you have to replace several blueberry plants after you fill up the hole. The only answer to this problem is a multi pronged attack using traps and poison. Traps are time consuming and often not very effective but with consistent determined use they can reduce you pocket gopher population. Poison is not perfectly effective either. You will need to place it in the gopher tunnels in early spring and late fall when the beasts are most active and hungry. Again, you will probably only thin them out and not eliminate them altogether. I use a hand punch that helps locate the tunnels and then dispense some treated grain. I don't love this method but it helps. I would like to have a small anhydrous ammonia dispenser to shoot some liquid ammonia into the burrow. I'm sure it would have a fine effect on the gopher but I haven't found a suitable small scale tank and I'm not sure it's legal.

Smoke bombs are available for placement in the gopher tunnels but I have not found them to be effective.

The third problem you will likely run into is trees. Not big ones - little ones. Remember, you are planting these blueberry plants for 30 to 45 years. Every bird that comes in to steal berries leaves a deposit containing seeds along with the fertilizer. The wind will blow tree seeds into your berry rows. These seeds will sprout and you won't notice them till they are a few years old and then you can't pull them out. If you cut them off they will sprout up again, and again, and again

If the weed trees hold their leaves until after the blueberry bushes have lost theirs you can bend the tree out and away from the blueberry plants and give it a good spray of 2,4,D. Alternatively, you can get a pistol syringe and actually inject the tree with a 20% to 30% solution of 2,4,D, Tordon, or other appropriate herbicide. To control weed trees with the pistol syringe you will need a stout hypodermic needle, I use a 14 gauge. I think it might be strong enough to inject rocks. To inject the 2,4,D solution, bend the tree down and slice the bark, with the tip of the syringe, for an inch or two. Then pry up a little pocket under the bark and inject 0.5 ml. of 30% 2,4,D solution into the pocket. Sometimes I just crack the tree off near the base and drop some 2,4,D solution onto the tree stub. While this system is not 100%

effective it will keep weed trees to a minimum with several application each year. You must be wearing heavy rubber gloves for this procedure as you don't want any contact with the 2,4,D solution.

Other annual and perennial weeds will be an ongoing problem. The usual farm weeds in your area will adjust to the acid soil in about 30 seconds. It may slow a few down a little but there are plenty of weeds that seem to enjoy the acidity. Foxtail, lambsquarters, milk weed, and velvet leaf are our main annuals. Good old quackgrass, and the trees are the worst perennials. There are several good pre-emergent herbicides that are effective on the annuals.

Simizine may be used at half the usual label rate on new plantings. Do wait until after you have gotten a good rain or have irrigated a few times before you apply, but don't wait till the weeds germinate. This will help with weed control, but it will not do the whole job. Those weeds that sprout through the herbicide should be cut off or mowed short. You can hoe them or pull them also but you will reduce the effectiveness of the herbicide if you break up the soil much.

I don't mean to imply here that strictly manual weed control is not effective. It is much easier on mature highbush blueberries where mowing and heavy mulch will accomplish 90% of the job.

Halfhigh blueberry varieties will require a lot more hand pulling and hoeing to keep the weeds at bay. It is relatively expensive when compared to proper use of chemical herbicides. If you have a couple of kids who need to earn some money, manual cultivation may work, but my situation has required the use of weed control chemicals. If you do use manual weed control remember that the blueberry roots are mostly just below the soil surface. If you cultivate too deeply you can easily injure the roots.

After the plants have been in the ground a year you should use a regular preemergent herbicide program. The following chemicals are in general use in the blueberry industry. Simizine and Diuron are broad spectrum and do a pretty good job on germinating annual grasses and broadleaf weeds. Norflurazan and Terbacil are also broad spectrum preemergents that have some action on young herbaceous perennial weed plants as well as annuals. General commercial practice includes tank mixing Simizine or Diuron with Norflurazan or Terbacil. The tank mix is changed each year so you can rotate through all the possible combinations in five or six years. This avoids buildup of any one chemical in the soil, and avoids the development of resistance to any one chemical in the weed population. Always read the label. Always follow label directions for rates of application on different soil types.

Spraying herbicides onto a heavy mulch will reduce the effectiveness of most preemergent herbicides at the same time it reduces the germination of most weed seeds. We try to spray early in the year to catch the weeds before they sprout and then we put a new layer of mulch down. This helps seal the preemergent herbicide into the soil so it can do it's job and gives a nice fresh layer of mulch for the pickers to work on later in the season.

BIG MISTAKE # 3: There is an old saying about putting a ten dollar tree in a twenty dollar hole. It's true - you should spend twice as much, or more, on site preparation as you spend on the plants going in the ground. We tried to cut a few corners at the start by not spending money on weed control prior to planting our first few acres of blueberries. We also didn't use a preemergent herbicide after we planted. We didn't mulch. Stupid, stupid stupid. We are still, after eight years, fighting perennial weeds in those plantings. They are productive now but it took an extra year or two for the bushes to mature. That has cost us money several ways. Extra work and chemicals for weed control, reduced yield, and the berries are harder to pick because of interfering weeds. Take care of weed control at the front end of the project. You will get to spend plenty of time and effort controlling weeds even if you do a good job at the start

Winter Care

Preparing blueberry plants for winter survival at the northern limits of blueberry culture may entail some extra work. Most commercial varieties of the Highbush blueberry are reliably hardy in USDA zone 5, with several varieties hardy into zone 4. The Halfhigh varieties are generally hardy into zone 3 (-35F).

The most important winter protection you can do is to purchase varieties that are winter hardy in your area. You will find, however, that there is not total agreement on the hardiness of each variety. Each supplier's catalog, or reference source will have slightly differing hardiness ratings for different varieties. This variability in hardiness ratings is based on different experiences, different research, and different microclimates which growers and researchers work with. I have seen varieties rated as generally not hardy (Berkeley and Jersey) produce fine crops every year in northwest Wisconsin, and varieties rated as very hardy (Rubel) rarely produce a decent crop in this area. When I pick varieties I usually try to "average out" the information that I have. I'll sit down with four or five catalogs and books and list out each source's hardiest varieties. If a certain variety only scores well from three out of five sources I'll move it to the questionable list.

The second most important winterization technique is to allow the plants to winterize themselves normally. Remember, **no nitrogen fertilizer** after the middle of June. Later nitrogen applications will cause late, soft shoot growth that will not have time to harden off, and will very likely winter kill. After the fruiting season blueberries need less water. About a half inch per week should be adequate on most soils, slightly more on very light sandy soils. Reduce irrigation to the minimum. Excess water in August can also cause late shoot growth and consequent winter kill of soft tissue.

I usually stop all weed control activities after the picking season is over. I have myself convinced that this is not just laziness. It reduces both nitrogen and water availability to the blueberry plants during the late part of the growing season. The dead weeds also help reduce wind and trap snow in the blueberry patch throughout the winter. On clean cultivated fields some growers seed ryegrass after harvest. This provides a winter snow trap and windbreak. In the spring it can be mowed and blown onto the rows as mulch.

Windbreaks help to reduce wind chill in the blueberry patch during the winter months. It is not simply absolute winter temperatures that cause winter kill. Low humidity and powerful drying winds desiccate plant tissue and often cause more damage than the coldest temperatures. Wind protection in the form of windbreaks around the patch, and tall weeds left in the patch will help reduce this problem.

In extreme situations fall spraying of anti-transpirents can reduce winter kill due to desiccation. Anti-transpirents are usually spray-on waxes that coat leaf tissues and reduce the amount of water the plants lose to the atmosphere. They are occasionally used in the summer during very hot dry weather to reduce irrigation requirements. They have some effect in winter to reduce desiccation injury.

If you are trying to raise blueberries under really extreme winter conditions, zone 3 (-30 to -40 degrees), you will need the hardiest "half high" varieties, and you will need to arrange to have them covered over the winter months. In areas where there is **reliable** snow fall, enough to cover the plants (up to three and one half feet tall), snow is the best insulator. This can work well in areas with "lake effect" snow. In areas with less snow the use of snow fence to trap blowing snow may work very well.

Remember that it will take a line of snow fence placed every 20 to 40 feet across the blueberry patch to provide the best coverage.

Where snow cover is not reliable, or in areas that get less than thirty inches of snow a year, you may still be able to winter your blueberries successfully under spun bonded, or floating, row cover. This is the material often seen in gardening catalogs to keep pests off plants or for stretching the growing season out as a frost cover. A single layer of the heaviest fabric, or a double layer of the lighter weights, used in conjunction with snow fence to trap available snow, will minimize winter injury to the plants. The winter cover should be placed over the plants about a month after they have gone dormant (lost their leaves), hopefully after some snow has fallen. The cover should be removed in early spring about when the ground has thawed.

Pruning

Pruning blueberry bushes usually starts when the bushes are five years old. On very vigorous bushes pruning may start a year earlier with the removal of spindly interior growth. The purpose of pruning is to balance berry production over time, promote ease of picking, allow air circulation and sun into the plant, and removal of dead and less vigorous wood.

The blueberry plant sends up a number of large canes from the crown of the plant or from the base of older canes each year. The next year these canes send out side branches, then branches are added to the existing side branches each year. These canes are most productive when 3 or 4 years old and slightly less productive in years 2 and 5. The major goal of pruning is to keep the population of canes on any one plant evenly divided between canes from 2 to 5 years old. All six year old canes should be removed. One or two extra one year old shoots should be left in case some get damaged and needs to be removed. The "perfect" mature blueberry plant will have 5 or 6 sturdy one year old canes, and 3 or 4 each of 2 to 5 year old canes. Actually, given variations in vigor for blueberry plants of different varieties and in different soil conditions, the pruning "ideal" will change in different situations.

Very vigorous blueberry plants may be allowed to carry an extra cane in each age group, while less vigorous plants should be pruned to fewer bearing canes. Heavy pruning may actually increase long term berry production from less vigorous plants by stimulating stronger new growth.

Berry size can, to some extent, be controlled by pruning. You can prune out up to 30% of the fruit buds in a given blueberry plant without affecting the total weight of berries harvested. You will get fewer berries but they will be larger and they will ripen earlier. This is important if you are selling berries U-pick. Large, flavorful berries are easier to pick and bring customers back.

The blueberry plant should also be pruned with an eye toward ease of picking. You don't want all the 5 year old canes in one place and the whole bush pruned so that the canes are too close together. On the other hand you do want to remove canes growing out into the row middles, floppy canes, or canes that are rubbing or crossing other canes. The "perfect" blueberry plant will be shaped like a vase or a flower pot, - thinner at the bottom and growing upward and outward in a neat and controlled manner. The canes of the plant should be distributed around the edge of the vase to a large extent, with the center left open. This shape serves several purposes. Most of the berries will be produced toward the outside of the plant making them accessible to the picker. Also, with an open center, there is more air circulation within the plant, and more sunlight in the interior of the plant. This promotes earlier ripening of interior berries, more even ripening of the entire crop and more production of flower buds. It also reduces the possibility of fungal diseases in the plant as it dries out sooner after rain and nightly dew.

Pruning is a skill that requires practice as much as knowledge. As you get to know your plants you will see what is needed. The general rules are (1) take out dead and diseased wood, (2) remove old and weak growth, (3) balance the age of the producing canes, (4) work toward an open balanced plant. Balance the need for a well pruned plant with the fact that you may have several thousand plants to prune. Remember that you will have years to get them exactly where you want them, that you probably never will get things perfect, and that good enough is good enough. Also remember that everyone will prune differently. Don't spend too much time arguing about it: good enough is good enough.

You will have piles of prunings left at the end of the winter. We tried leaving them in the row middles on the theory that they would get chopped up by the mower when we mowed the grass in the middles. This didn't work very well - there were a lot that didn't get chopped up very well and they were underfoot during picking season. A better solution is to pile the prunings on a tarp and drag them off the field to be burned or chopped. We usually drag our prunings to an area where we need some mulch for erosion control. Come spring we get out our bale chopper/mulcher and turn the prunings into shreds and small pieces. Burning is probably best in commercial growing area as it allows for destruction of disease organisms in the pruned material.

Marketing

You should start planning you marketing strategy before you put a blueberry plant in the ground. As a small grower you cannot compete with the large commercial growers who sell blueberries for processing at 45 cents a pound. In areas where blueberries are common even U-Pick prices don't exceed 90 cents a pound. At U-Picks in areas where blueberries are not common you can set your price. As a rule of thumb I set our price per pound at or a little below the local retail price of a pint (about 10 to12 ounces) of fresh blueberries at the grocery store a week or two before the season starts. That way I am below the retail price with far superior quality.

I always explain that the customers get only the berries they want to pick, no soft or moldy berries in the bottom of the box, and they can have free samples as long as they are picking. Often, just as our season opens, Michigan blueberries are in the stores as "loss leaders" for little more than the cost packing and shipping - sometimes they give them away. Don't try to compete on price, compete on quality. If anybody mentions giveaway berries I just make the following offer: "Bring a box of the free berries out here and pick a box of our berries and do your own taste test. If you can look me in the eye and say the ones from the store are just as good as the ones you picked here you can have them both free." It doesn't always bring the customer out but it does convey an image of quality and that will get around.

If you are planning on a U-pick operation you must have reasonable proximity to one or several large population centers. If your have a scenic site or are in a well traveled tourist area (Rush River Produce is both) you can stretch "reasonable proximity" out to 60 miles or more. If you have a more "run of the mill" site for your blueberry patch you may have a smaller market area. 20 miles is about the average distance people will travel to pick berries, though some studies indicate people will travel up to 40 miles for blueberries. At Rush River Produce 65% of our customers travel 60 miles, but we are lucky to have a very attractive site in a well traveled tourist area. If your are planning to sell your berries at a farmers market or through local stores, your location depends on how far you are willing to drive to your market. I strongly suggest that you start planning your marketing strategy at the beginning, or before you start planting. Plan early but be willing to change your direction if something is not working.

Diversity is another key to good marketing. Customers may come out to buy blueberries but they always have their eye out for something else that is interesting. We also have garlic, black and red currants, apples, raspberries, and we are adding gooseberries, and lingonberries in the next few years. This diversity gives the customers something else to buy and something to try. It doesn't matter if you don't make a lot of profit on these extras, they will help bring people back and give them something to talk about to their friends, who may ask them for directions. It also gives you something to give away as a treat or incentive.

We make it a practice to give our customers something extra as often as possible. We can't always do it on real busy days, but if I hear somebody say "what's that?" I offer to let them pick a pint or two to try. Even if they don't like the produce they will appreciate the offer. If they do like it I've got a new customer for that produce. It's a classic win/win situation. I may be out a little produce in the short run, but I've got people who never tasted black currants coming back for more, and people who didn't know you could grow lingonberries in this area are anticipating making lingonberry jam in the next year or so.

The U-Pick experience of the customer is as important, in some ways, as the berries you are selling. There is a lot of talk about Agricultural Tourism these days in our area. We find that we are selling a day in the country for the family and we encourage this. We encourage bringing children out to pick berries (many U-Picks discourage them). We have a play area where customer kids can play with our kids. We have picnic tables, and we sell pop and a few treats. We want to encourage family outings and our customers respond well to the encouragement.

We have also spent a great deal of time and effort planting perennial flowers and landscaping the areas that people walk through the most. Many of our customers have suggested that we could charge admission just to walk around and look at the flowers. We don't want to do that but we do want to give customers plenty of reasons to come back.

Press Coverage

Your local media: TV, radio, newspapers, shoppers, travel magazines, etc. are very important tools to develop your market. Believe me: they **do** want to write stories about what you are doing. There are a few things that you need to do to get your name and your berry growing business a lot of excellent press coverage in your area.

1. **You need to let them know that you are there.** This must be more than the occasional chance meeting or random phone call. As a business person you must start a regular and persistent program of contacting all the media outlets in your chosen market area and beyond. You will need to develop a complete list, with contact names, of every newsletter, shopper, newspaper, radio station, TV station, and magazine in your market area. Don't be too choosy. Include everything you can find. Contact

local Chamber of Commerce offices, check the yellow pages in phone books, look for tourist oriented publications in shops and tourist stops. We got an excellent list of over 200 media outlets in 90 mile radius of our county from our county economic development office.

We added a few outlets that we advertised in that weren't on the list. Then we started cutting some of the media outlets that probably wouldn't be interested in us (Law Review, Building Contractors Journal, etc.) or were very local and out of our marketing area. We cut our list to 150 media outlets.

In April of 1995, with snow still on the ground, we decided to mail out a press release to every media outlet on our list every month until the end of our berry season.

2. **Tell them how to contact you.** If you mail a letter to your local newspaper, or give them a call, they will read it and may respond. Local newspapers especially are interested in new and interesting local businesses.

This past year we decided that we would get serious about press releases. If you want to get to a wider audience in larger towns you have to get in that media. To get to the larger media outlets you need to follow the rules of the game. A letter about your business to a large newspaper will probably get read and tossed in the waste basket. A press release, in the proper format, will get read and probably put in a file and may be referred to later for story ideas.

A press release must answer the questions who, what, where, when and why. It is vehicle to make it easy to find you and your business, have your name and phone number prominently displayed. It must have a release date and an end date. Your business address and phone number should be easily visible.

A sample press release format may look like this:

Contact:
John or Terry Cuddy
715 594-3648

Release Date: 7/8/95
End Date: 8/1/95

Lead Information:
Who: Terry and John Cuddy
What: Fresh blueberries, raspberries and garden produce.
Where:Rush River Produce, Maiden Rock, WI
When: July, August, and September
Why: The freshest, highest quality produce in the area.

Rush River Produce
W4098 200th Ave
Maiden Rock WI 54750
715 594-3648

The Blueberries are ripe and the picking season has started. Rush River Produce, the area's largest U-pick blueberry farm, is ready for the fruits of the season. At Rush River Produce folks pick their own berries and have a fun day in the country. Kids of all ages love the taste of fresh fruit and you can eat your fill while you pick some to take home to enjoy later. Seven acres of blueberries ensure that there will be plenty of ripe, delicious blueberries for everyone.

Blueberry season begins July 8 and will last until early August, followed by red ripe raspberries in August and September. Along the way through the summer months you will find black currants, apricots, locally produced honey and maple syrup, and garden produce available to those who venture up the bluffs to the farm overlooking Lake Pepin and the Rush River Valley.

Rush River Produce is located just 3 miles off Highway 35, Wisconsin's' Great River Road, outside of Maiden Rock, WI. Please call (715) 594-3648 for picking information and directions. The berry farm is a scenic one hour drive from the Twin Cities, Eau Claire, or Rochester areas. For your convenience, they are open seven days a week, 8 a.m. until dark. "We provide the boxes, you take home the harvest."

Your press releases should be kept to a single page and double spaced. You want to present the germ of a story to the editor or reporter who reads the release.

3. **Tell them why they should contact you and write a story about you berry farm**. The more interesting the press release is, the more probable a response will be, but don't think that a press release has to be gripping, exciting prose. Remember, you should be planning to contact these media outlets several times over the course of the year. You won't produce a great press release every time, you don't need to.

Your press releases may be about topics that you are interested in, projects you are doing on the farm, or new crops you are trying out for the future. Of course you must talk about blueberries in each press release. We always mention that we have the best U-pick blueberries in the area. But our most successful press release was about growing Lingonberries. That press release probably got us as much free press space as all the other press releases put together. All that interest in lingonberries brought us a lot of blueberry customers.

Now, as soon as I get an idea for a new crop to try, I start writing the press release in my mind. Next year we are going to have our first crop of Gooseberries. A lot of the people who read these press release may not even know what a gooseberry is. A few of them will call us to find out what they are and what to do with them. If we manage to communicate enthusiasm and interest to the reporter the odds are that they will come out to do an in depth story. Perfect!

This approach to press releases should get you better known in your market area. Even if you only "hit" about a 5% success rate it will mean a lot of free ink for you. It should also push you into trying new projects just to be interesting to the media.

Conclusion

The information presented in this book represents the state of our experience combined with what we have learned from other growers, publications, and researchers. Since every soil and micro climate is different what works for us may not work as well, or at all, for you. We would like to update this publication as often as we can. If you have suggestions for information not included in this book, tips or techniques that have worked for you (or haven't worked for you), or any other comments, please send then to us so we can improve this publication. Unless otherwise requested, sources for really good ideas will be mentioned by name in the publication.

Questions ? Comments ?
Suggestions ......
Call us at:
715 594-3648

Appendix A: Technical Publications

Blueberry Science
Eck, P
Rutgers University Press
New Brunswick, New Jersey

The Blueberry Bulletin
Dr. Gary C. Pavlis, County Agricultural Agent
6260 Old Harding Way
Mays Landing, NJ 08330
$10/yr, Weekly during growing season

Highbush Blueberry Management
Bob Gough (HB)$44.95 (PB)$24.95

Highbush Blueberry Production Guide
Northeast Agricultural Engineering Cooperative Extension.
(3 Ring Binder)$45.00

Blueberry Culture
Eck and Childers
(HB) $40.00

Blueberry Science
Eck
(HB) $49.00

Small Fruit Crop Management
Himmelrich and Galetta
$59.00

Northland Berry News
2124 University Avenue W.
St. Paul, MN 55114-1838 ($15/yr Sub)

Great Lakes Fruit Grower News
343 S. Union St.
Sparta, MI. 49345 ($18/3 yr. Sub)

Technical Publications, Other;

Economics for Small Scale Food Producers
Christensen, Robert L.
University of Massachusetts
Amherst, Mass. 01003

Your Money or Your Life
Dominguez and Robin
Viking Press

Guerrilla Marketing
Levinson, Jay C.
Houghton Mifflin

Farming Alternatives
N R A E S
152 Riley Robb Hall
Cornell University
Ithaca, NY 14853
(Ask for their publications catalog too!)

Sell What You Sow
Great Lakes Publishing
PO Box 128
Sparta, MI 49345 ($25)

Appendix B: Commercial Blueberry Suppliers

Barwacz's Farm
05146 59th St.
Grand Junction, MI 49506
(616) 253-4419
Fax (616) 253-4495

Boston Mountain Nursery
20189 N Hwy 71
Mountainburg, ARK 72946
Phone/Fax: (501) 369-2007

DeGrandchamp's Nursery
15575 77th St.
South Haven, Michigan, 49090
phone: (616)637-3915
Fax: (616)637-2531

Hartmann's Plantations
310 60th St., PO Box E
Grand Junction, Michigan, 49056
Phone: (616) 253-4281
Fax: (616) 253 4457

Indiana Berry and Plant Co.
5218 West 500 South
Huntingburg, IN 47542
Phone 1-800-295-2226

North Star Gardens
19060 Manning Trail North
Marine on St. Croix, MN 55047-9723
Phone: (612) 227-9842
Fax: (612) 227-9813

Nourse Farms
41 River Rd
South Deerfield, MA 01373
Phone: (413) 665-2658

Pense Nursery
16518 Marie Lane
Mountainburg, AR 72946
Phone:/Fax: (501) 369-2494

Tower View Nursery
70912 CR-388
South Haven, Michigan, 49090
Phone: (616) 637-1279
Fax: (616) 637-6257

Appendix C: Other Resources:

Your County Extension Agent
at your county courthouse.

The following States have excellent publications on blueberry culture and management:

Arkansas
Florida
Georgia
Louisiana
Kentucky
Maine
Massachusetts
Michigan
Minnesota
New Jersey
North Carolina
Oregon
Pennsylvania
South Carolina

Texas
Washington

Your State Fruit Growers Association

Your State Tourism Department

Other Berry Growers in your area

Additional copies of this book are available from:

Rush River Publications
W4098 200th Ave.
Maiden Rock, WI 54750

Please Send $18.50 for your post paid copy of

Growing Blueberries
A Guide For The Small Commercial Grower